Printed Batteries

Printed Batteries

Materials, Technologies and Applications

Edited by

Senentxu Lanceros-Méndez

BCMaterials, Basque Center for Materials, Applications and Nanostructures, Spain and Center of Physics, University of Minho, Gualtar campus, Braga, Portugal

and

Carlos Miguel Costa

Centers of Physics and Chemistry, University of Minho, Gualtar campus, Braga, Portugal

Registered Office(s)
John Wiley & Sons, Inc., 111 River Street, Hoboken, NJ 07030, USA
John Wiley & Sons Ltd, The Atrium, Southern Gate, Chichester, West Sussex, PO19 8SQ, UK

Editorial Office
The Atrium, Southern Gate, Chichester, West Sussex, PO19 8SQ, UK

For details of our global editorial offices, customer services, and more information about Wiley products visit us at www.wiley.com.

Wiley also publishes its books in a variety of electronic formats and by print-on-demand. Some content that appears in standard print versions of this book may not be available in other formats.

Library of Congress Cataloging-in-Publication Data

Names: Lanceros-Méndez, Senentxu, 1968– editor. | Costa, Carlos Miguel, 1991– editor.
Title: Printed batteries : materials, technologies and applications / edited by Senentxu
    Lanceros-Méndez, Carlos Miguel Costa.
Description: Hoboken, NJ : John Wiley & Sons, 2018. | Includes bibliographical
    references and index. |
Identifiers: LCCN 2017054470 (print) | LCCN 2018000676 (ebook) | ISBN 9781119287889 (pdf) |
    ISBN 9781119287896 (epub) | ISBN 9781119287421 (cloth)
Subjects: LCSH: Electric batteries. | Three-dimensional printing.
Classification: LCC TK2896 (ebook) | LCC TK2896 .P755 2018 (print) | DDC 621.31/2424–dc23
LC record available at https://lccn.loc.gov/2017054470

Cover design by Wiley
Cover image: © D3Damon/Getty Images

Set in 10/12pt Warnock by SPi Global, Pondicherry, India

Printed in Singapore by C.O.S. Printers Pte Ltd

10 9 8 7 6 5 4 3 2 1

Contents

List of Contributors

Ana Claudia Arias
Electrical Engineering and Computer
Sciences Department
University of California, Berkeley
USA

Reinhard Baumann
Department of Printed
Functionalities
Fraunhofer ENAS
Chemnitz
Germany
and
Department of Digital Printing and
Imaging Technology
Chemnitz University of Technology
Germany

Li-Chun Chen
Department of Chemical Engineering
National Tsing Hua University
Hsinchu
Taiwan
and
Material and Chemical Research
Laboratories
Industrial Technology Research
Institute
Hsinchu
Taiwan

Keun-Ho Choi
Department of Energy Engineering
School of Energy and Chemical
Engineering
Ulsan National Institute of Science
and Technology (UNIST)
Korea

Maurice Clair
3D- Micromac AG
Chemnitz
Germany

Carlos Miguel Costa
Centers of Physics and Chemistry
University of Minho
Gualtar campus
Braga
Portugal

Abhinav M. Gaikwad
Electrical Engineering and Computer
Sciences Department
University of California, Berkeley
USA

Diana Golodnitsky
School of Chemistry and Applied
Materials
Tel Aviv University
Israel

Martin Krebs
VARTA Microbattery GmbH
Innovative Projects
Ellwangen
Germany

Senentxu Lanceros-Méndez
BCMaterials
Basque Center for Materials
Applications and Nanostructures
Spain
and
Center of Physics
University of Minho
Gualtar campus
Braga
Portugal

Sang-Young Lee
Department of Energy Engineering
School of Energy and Chemical
Engineering
Ulsan National Institute of Science
and Technology (UNIST)
Korea

Darjen Liu
Department of Chemical Engineering
National Tsing Hua University
Hsinchu
Taiwan
and
Material and Chemical Research
Laboratories
Industrial Technology Research
Institute
Hsinchu
Taiwan

Ta-Jo Liu
Department of Chemical Engineering
National Tsing Hua University
Hsinchu
Taiwan

Svetlana Menkin
School of Chemistry
Tel Aviv University
Israel

Kalyan Yoti Mitra
Department of Digital Printing and
Imaging Technology
Chemnitz University of Technology
Germany

Juliana Oliveira
Center of Physics
University of Minho
Gualtar campus
Braga
Portugal

Aminy E. Ostfeld
Electrical Engineering and Computer
Sciences Department
University of California, Berkeley
USA

Patrick Rassek
Hochschule der Medien (HdM)
Innovative Applications of the
Printing Technologies (IAF/IAD)
Stuttgart Media University
Germany

Ela Strauss
Ministry of Science, Space
and Technology
Jerusalem
Israel

Carlos Tiu
Department of Chemical Engineering
Monash University
Clayton
Australia

Anh-Tuan Tran-Le
Department of Digital Printing and
Imaging Technology
Chemnitz University of Technology
Germany

Michael Wendler
ELMERIC GmbH
Rangendingen
Germany

Andreas Willert
Department of Printed Functionalities
Fraunhofer ENAS
Chemnitz
Germany

Preface & Acknowledgements

He who sees things grow from the beginning
will have the best view of them.

Aristotle (384 BC–c. 322 BC)

Printed batteries are an excellent alternative to conventional batteries for an increasing number of applications such as radio frequency sensing, interactive packaging, medical devices, sensors, and related consumer products. These batteries result from the combination of conventional battery technologies and printing technologies. Printed batteries are increasingly being explored for highly innovative energy storage systems, offering the possibility for better integration into devices and novel application areas.

In this context, the main motivation of the present book is to offer the first comprehensive account on this interesting and growing research field providing the main definitions, the present state of the art, the main research issues and challenges, and the main application areas. In this scope, this book summarizes the frontline research in this fascinating field of study, presented by selected authors with truly innovative and preponderant work.

The book provides an introduction to printed batteries and the current state of the art on the different types and materials, as well as the printing techniques for these batteries. Further, the main applications that are being developed for those printed batteries are addressed as well as the principal advantages and remaining challenges in this research field.

The first chapter provides a general overview of the area of printed batteries. It deals with definitions and the main printed batteries types such as lithium-ion, Zn/MnO_2 and related systems. The advantages and disadvantages of printed batteries are discussed and the main applications summarized. Chapter 2 describes the printing techniques used for the production of printed batteries and gives a brief description of materials, substrates and the process chain used in printed batteries. Chapter 3 deals with the important issue of the influence of slurry rheology on electrode processing through its formulation, preparation technique, coating and drying systems. Moreover, the rheological characteristics of the electrode slurry are described.

Chapter 4 focuses on the polymer electrolytes used for the development of printed batteries. The state of the art on polymer electrolytes produced with different printing techniques is described in this chapter, as well as the electrolytes used in conventional and lithium-ion batteries.

The subject of Chapter 5 is the design of printed battery components. This chapter focuses on printed material layers for the electrodes used in Zn/MnO_2 batteries, lithium-ion batteries, and related systems.

Chapter 6 presents the main applications of printed batteries. Power electronics, RFID, sensors and actuators, medical and energy-harvesting devices are presented and discussed.

Taking into account the different applications of printed batteries, Chapter 7 provides an industrial perspective on printed batteries considering relevant industrial aspects such as layout considerations, current collectors, carrier substrates and multifunctional substrates, among other topics.

Finally, Chapter 8 summarizes some of the main open questions and challenges and the outlook for this research field.

This book would have not been possible without the dedicated and insightful work of the authors of the different chapters. The editors truly thank them for agreeing to devote their precious time to this enterprise. We thank them for their kindness, dedication and excellence in providing high-quality chapters illustrating the main features, challenges and potential of the area of printed batteries. It has been a pleasure and an honor to work with you in this important landmark in the field!

Additionally, this book would not have been possible without the continuous dedication, support and understanding of our research group colleagues both at the Center of Physics, University of Minho, Portugal, and the BCMaterials, Basque Center for Materials, Applications and Nanostructures, Leioa, Spain. Thank you all for the beautiful and continuous endeavor of driving science and technology a step further together and for sharing this important part of our lives!

Last but not least, we truly thank the team from Wiley for their excellent support: from the first contacts with Rebecca Ralf and Sarah Higginbotham to the last with Shagun Chaudhary, Máire O'Dwyer, Emma Strickland, Rajitha Selvarajan and Lesley Jebaraj, passing through the different colleagues that supported this work; your kindness, patience, continuous support, technical expertise and insights were essential to make this book come true. It has been a real pleasure to work together with you!

Finally, let us hope this first book on printed batteries will promote not only a deeper understanding of this increasingly relevant research and application area but also the interest and motivation to tackle the main challenges, so that we all together contribute to a bright and innovative future in the area of printed batteries!

Carlos Miguel Costa and Senentxu Lanceros-Méndez

1

Printed Batteries: An Overview

Juliana Oliveira[1], Carlos Miguel Costa[1,2] and Senentxu Lanceros-Méndez[1,3]

[1] Center of Physics, University of Minho, Gualtar campus, Braga, Portugal
[2] Center of Chemistry, University of Minho, Gualtar campus, Braga, Portugal
[3] BCMaterials, Basque Center for Materials, Applications and Nanostructures, Spain

1.1   Introduction

Increasing technological development leads to the question of how to efficiently store energy for devices in the fields of mobile applications and transport that need power supply [1, 2]. Energy storage is thus not only essential but also one of the main challenges that it is necessary to solve in this century [2, 3].

Further, energy storage systems are also increasingly needed, among others, to suitably manage the energy generated by environmentally friendly energy sources, such as photovoltaic, wind and geothermal [4, 5].

Batteries are the most-used energy storage systems for powering portable electronic devices due to the larger amounts of energy stored in comparison to related systems [2, 6]. Among them, the most widely used battery type is lithium-ion batteries, with a market share of 75% [7].

Anode, cathode and separator/electrolyte are the basic components of a battery, the cathode (positive electrode) being responsible for the cell capacity and cycle life. The anode (negative electrode) should show a low potential in order to provide a high cell voltage with the cathode [8–10].

The separator/electrolyte is placed between the electrodes as a medium for the transfer of lithium ions and also to control the number of lithium ions and their mobility [11].

Advances in the area of batteries in relation to printed technologies is expected to have a large impact in the growing area of small portable and wearable electronic devices for applications such as smart cards, RFID tags, remote

Printed Batteries: Materials, Technologies and Applications, First Edition.
Edited by Senentxu Lanceros-Méndez and Carlos Miguel Costa.
© 2018 John Wiley & Sons Ltd. Published 2018 by John Wiley & Sons Ltd.

sensors and medical devices, among others. This in fact originated in the development and proliferation of smart and functional materials and microelectromechanical systems (MEMS) needing on-board power supply to provide capacities of 5 to 10 mAh.cm^{-2} with overall dimension of < 10 mm^3 [12–14].

The technological advances of the past years and the need for low-cost and simple processing leads to the potential replacement, in some areas, of conventional processing technologies by printed technologies, as evidenced in applications such as sensors, light-emitting devices, transistors (TFT), photodiodes, flat panel display solar cells and batteries, among others. Printed technology characteristics such as low cost, large area, high volume, light weight, and the processing of multilayered functional structures on rugged and flexible substrates, pave the way for new production paradigms for specific application areas [15–17].

In fact, it is expected that the global market for printed electronics will reach $45 billion in 2017 and is estimated to exceed $300 billion over the next 20 years [15, 16, 18].

This fact is also evidenced by the many articles published in scientific journal about inks and printed electronics, as shown in Figure 1.1.

Printed materials for electronics can be applied on different substrates such as paper, plastics and textiles, giving origin to the term "flexible electronics". Typically, the most frequently used printing techniques for printed electronics are ink-jet and screen-printing [19], but related cost-efficient and high-throughput production techniques such as solution-processing techniques

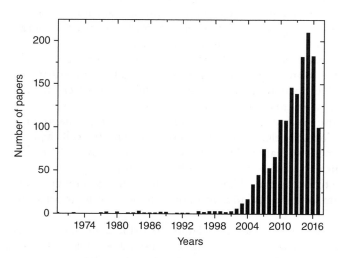

Figure 1.1 Research articles published related to inks and printed electronics. Search performed in Scopus database with the keywords "inks" and "printed electronics" on 19 June 2017.

including spin, spray, dip, blade and slot-die have been used, as well as gravure, flexographic and offset printing technologies [20, 21].

The different printing techniques require the use of specific inks with accurate control of viscosity and surface tension, among other things [22, 23]. Further, for specific printing techniques, the ink properties should be adjusted taking into account the specific pattern to be printed [24].

Printed electronics requires the use of different types of inks such as dielectric, semi-conductive or conductive, which are used to print the different active layers of the devices. Further, inks with piezoelectric [25], piezoresistive [26], and photosensitive [27] properties, among others, have been developed for the fabrication of sensor devices. Typically, inks can be defined as colloidal solutions as the result of a dispersion of organic and/or inorganic particles with specific size into a polymer solution [28]. Moreover, these inks must be cheap, reliable, safe to human health, and processable at temperatures below 50 °C. Further, the inks should preferentially show mechanical robustness, flexibility and recyclability [29].

Independent of the printing process, the ink should be distributed on the substrate with a specific pattern in a reproducible way, which strongly depends on its rheological properties [30].

The rheological properties (flow behavior, flow time and tack) of the ink can be evaluated by using the rotational viscosimeter to measure the viscosity as a function of shear rate, as the material is subjected to multiple shear rates during material processing.

In particular, it is important to prevent the agglomeration or sedimentation of the particles through attractive/repulsive forces, which depends on processing shear rate, as this will strongly affect the final properties of the printed layer [31].

At low shear rate, the viscosity of the inks is higher due to the attraction between particles, which induces their flocculation and immobility. At higher shear rates, the viscosity of the inks decreases through the low flocculation and higher mobility of solvent entrapped between particles [32, 33]. However, the viscosity of printing inks is not only a function of the shear stress but also of time, which plays an important role in the flow process of the ink for each printed element [30].

Further, the physical and chemical stability of the inks is affected by the different fabrication steps (stirring, dispersion, etc.), in which the energy input and mixing time influence both particle stability and degree of dispersion [34].

The combination of printing and battery technologies gives rise to printed batteries; for this at least one of the components should be processed and deposited through printing techniques in order to keep that designation [12, 35].

Figure 1.2 shows the origin of the denomination and the main applications of printed batteries.

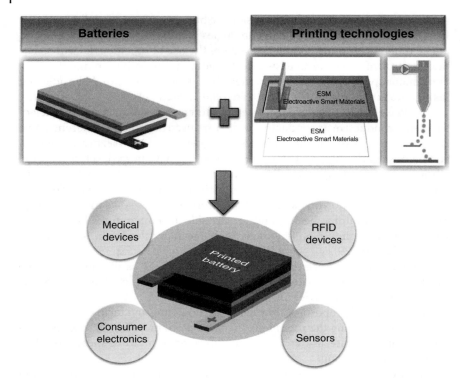

Figure 1.2 An overview of printed batteries and main applications. (*See insert for color representation of the figure.*)

Further, flexible/stretchable batteries [36, 37] and solid-state microbatteries [38] can be included within the printed battery area when one or more components are produced by printing technologies. In addition, there are usually non-printed components such as the current collector, which also serves as support for the printed structure.

Inks for printed batteries are typically composed of a polymer binder, a solvent and suitable fillers, depending on the layer type: electrodes and separator/electrolyte [35]. Suitable fillers are in the form of micro/nanoparticles, nanoplates, nanowires, carbonaceous matter or ionic liquid, among others [29]. The proper transfer of the ink from the printing plate to the substrate is the main function of a printing process [30].

In the field of printed batteries, ink rheology is one of the key issues, due to the high active material loading that may be necessary for proper battery performance. This ink rheology depends mainly on particle size, solid loading concentration and solvent type [39, 40], with adequate ink showing moderate viscosity and weak sedimentation behavior resulting in an homogeneous particle system within a polymer network [31].

The main printed battery component is the electrode (anode and cathode) [22], and different inks have been reported in the literature based on different active materials such as lithium cobalt oxide ($LiCoO_2$) [41] and lithium iron phosphate ($LiFePO_4$) [40] for the cathode, and graphite [42], mesocarbon microbeads (MCMBs) [43] and tin oxide (SnO_2) for the anode [44]. The active material content of the electrode affects its thickness, which in turn influences battery capacity: increasing electrode thickness leads to mass transport limitations of lithium ions in the electrolyte phase leading to a reduction in the capacity of the cell [45, 46]. Also the porosity of the electrodes has a strong impact on battery performance as it influences the effective electronic and ionic conductivity values [47].

On the other hand, the separator/electrolyte has not been printed very often due to the necessary low ionic conductivity, which leads to the use of composite gel electrolytes to achieve ionic conductivity values closer to those of conventional electrolytes [35]. The separator/electrolyte component of printed batteries is mainly based on composite gel electrolytes where the separator layer is soaked in an organic liquid electrolyte (salt dissolved into an organic solvent or ionic liquid to produce an ion-conducting solution in an inert porous polymeric membrane) in which it is important to control the swelling process [35, 48, 49].

Thus, one of the largest challenges is the development of inks for printing solid-state separator/electrolytes with a minimum ionic conductivity of 10^{-4} S/cm and mechanical and thermal stabilities [50].

The efforts and challenges involved in developing and optimizing specific inks for the different battery components that meet the requirements of efficiency, stability and processability for different printed techniques (Figure 1.3) are the main focus of the present fundamental and applied research efforts in this field.

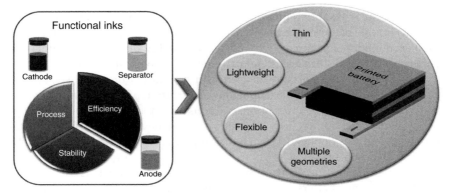

Figure 1.3 An overview of the functional inks and relevant requirements in the area of printed battery research. (*See insert for color representation of the figure.*)

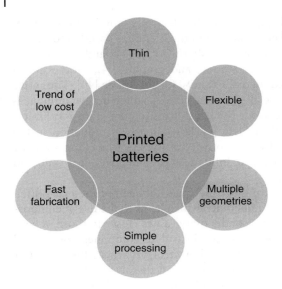

Figure 1.4 Main features and attributes of printed batteries.

The key features and attributes of printed batteries are that they are: customizable, thin, high power, low cost, mechanically flexible, lightweight and rechargeable and that they allow large printed areas. These features will allow the fabrication of functional systems with batteries already integrated in devices [51].

These features and attributes are shown in Figure 1.4 and are the main advantages in comparison to conventional batteries.

The production costs and processing steps for printed batteries can be reduced through the use of roll-to-roll production methods, as they enable the fabrication and assembly of the different layers of the batteries at high speed in a continuous process [52].

Some of the main differentiating factors of printed batteries in comparison to conventional batteries are their simple integration into devices, the possibility of production for large areas and the possibility of thickness reduction. Further, eco-friendly processes and materials are also possible [53].

Currently, research efforts are focused on improving performance (specifically energy and power) and on developing new fabrication processes, inks, designs and characteristics for applications such as smart cards, radio-frequency-identification (RFID) security and information devices, thin-film medical products, and new applications including e-labels, e-packaging, e-posters and medical disposables [54].

A common factor for many of the aforementioned products is the requirement of on-board battery power supply at the microwatt-level with specific designs, which can be achieved with printed batteries.

The performance parameters (i.e., power and energy value, lifetime and discharge rate) of printed batteries are established according to the application, the delivered capacity being between 0.7 and 90 mAh for commercial printed batteries [35, 55, 56].

Thus, printed batteries are being applied in an increasing number of applications and the advances in printable ink formulations and printing technologies, such as 3D-printing, will allow the fabrication of fully printed batteries with high areal energy density to widen the range of possible application areas.

1.2 Types of Printed Batteries

Electrochemical power sources, defined as batteries, were invented by Alessandro Volta, professor at the University of Pavia, Italy, in 1800, and are nowadays an essential component of the electronic devices market as well as of hybrid electric vehicles (HEVs) and electric vehicles (EVs) [57]. The "voltaic pile" consists of an alternating sequence of two different metal discs (zinc, Zn, and silver, Ag) separated by a cloth soaked in a sodium chloride solution [57]. Over the years, different battery types, including Zn, nickel-cadmium and nickel-metal hydride, were developed, with lithium-ion batteries now the most advantageous type.

The first Li-ion batteries were commercialized by Sony in 1991 based on the pioneering work of Yazami regarding the use of lithium-graphite as a negative electrode [58].

Some of the main advantages of lithium-ion batteries include being light, cheap, environmentally friendlier and safer as well as showing higher energy density, less self-discharge, no memory effect, prolonged service-life and higher number of charge/discharge cycles [9, 57].

Batteries are usually defined as primary and secondary, the latter being rechargeable batteries [59, 60]. Independently of the battery type, their main constituents are the two electrodes, anode and cathode, and the separator/electrolyte, as shown in Figure 1.5.

The two main processes of rechargeable batteries are charging and discharging, as illustrated in Figure 1.5. During the charging process, the movement of ions is from the cathode to the anode electrode and during the discharge, the movement is in the opposite direction, i.e., from the anode to the cathode [60].

In Figure 1.5, graphite represents the anode material and lithium-manganese oxide, $LiMn_2O_4$, represents the cathode material.

With respect to printed batteries, the most frequently used ones are based on lithium, Li, and zinc, Zn.

Lithium printed batteries are lithium-ion with different electrodes (graphite or Li_xC_n for anode and $LiCoO_2$, $LiMnO_2$, or $LiFePO_4$ for cathode, lithium-manganese dioxide, $Li-MnO_2$ and post Li, i.e., lithium-air, sulphur-cathode, etc.)[53].

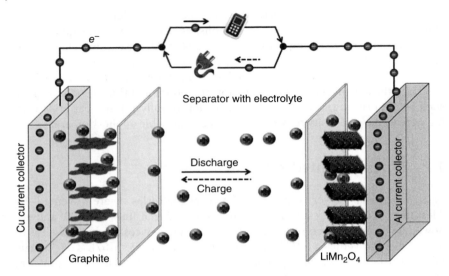

Figure 1.5 Schematic illustration of the main constituents and representation of the charge and discharge modes of a battery. (*See insert for color representation of the figure.*)

In relation to zinc batteries, the most frequently used are zinc-manganese dioxide, Zn-MnO$_2$ (Zn for anode and MnO$_2$ for cathode), zinc-air and zinc-silver oxide, Zn-Ag$_2$O [53]. Further, there are other electrochemical systems, such as nickel/metal hydride, which have been also applied in printed batteries [61].

As an example, for anodes based on graphite and cathodes based on LiCoO$_2$, the rechargeable electrochemical reaction of a lithium-ion printed battery system is:

$$LiCoO_2 + C_6 \xrightarrow[\text{dischærge}]{\text{chærge}} Li_{1-x}CoO_2 + Li_x C_6 \tag{1}$$

Typically, zinc battery types are non-rechargeable systems except for nickel-metal hydride. For the Zn-MnO$_2$ system, which is the most often used in printed batteries, the electrochemical reaction is:

$$Zn + 2MnO_2 + H_2O \rightarrow ZnO + 2MnO(OH) \tag{2}$$

For many small device applications in which no high voltage is required, Zn-MnO$_2$ batteries can be the most appropriate due to their high energy content, lower internal resistance, large shelf-life and the low cost of Zn and MnO$_2$ in comparison with lithium battery materials [62].

In relation to the other printed battery types, those that stand out in the literature after Zn-Mn$_2$O and lithium batteries are Zn-Ag$_2$O batteries with 1.3 to 5.4 mAh.cm^{-2} at 1.5 V [63, 64].

1.3 Design of Printed Batteries

For conventional batteries, there are basically four main designs, which are coin cell, prismatic, spiral wound and cylindrical [65]. All of them have in common the fact of being rigid and bulky, and not adequate for flexible electronics devices.

One advantage of the use of printing technologies in the fabrication of batteries is that it is possible to develop one or more layers with a specific pattern, i.e., design [66]. This is particularly relevant as together with the characteristics of the materials used for the fabrication of a battery, the geometry/architecture of the battery strongly affects its performance [67]. For printed batteries, the main types are the stack or sandwich architecture (Figure 1.6) and the coplanar or parallel architecture (Figure 1.7).

Figure 1.6 shows the stack or sandwich architecture, which consists of a current collector for the anode, anode, separator with electrolyte, cathode, and current collector for the cathode, all deposited in a flexible substrate with an overall thickness of 0.5 mm for the printed battery.

This architecture is identical to that of conventional batteries; it leads to low internal resistance due to the small distance that the lithium ions travel by moving between the anode and cathode, which also allows shorter charging times.

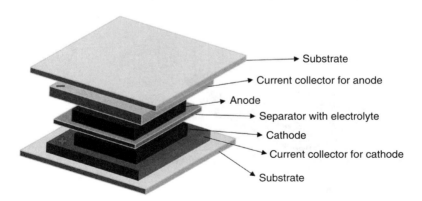

Figure 1.6 Schematic representation of a printed battery in the stack or sandwich cell architecture.

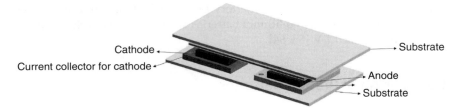

Figure 1.7 Schematic representation of a printed battery in the coplanar or parallel cell architecture.

Figure 1.7 shows the coplanar or parallel architecture for printed batteries, which consists of the anode and the cathode in a side-by-side position. This architecture is the most frequently used for stretchable batteries [68, 69]. In this geometry it may be not necessary to put the separator within the cell [53].

In this architecture, the risk of shorting during battery mechanical stretching is minimal.

It should be noted that, independently of the battery architecture, the sealing process is an essential step in printed batteries. This process consists of a sealing layer based on a polymer glue, which can be processed by the application of heat or pressure, with the main objective of protecting the battery against atmospheric gas molecules such as H_2O and O_2 [53, 70].

Energy storage within the sandwich architecture (Figure 1.6) has been increased by the development of interdigitated architectures, such as the one represented in Figure 1.8 [71]. The interdigitated architecture is based on electrode digits separated by an electrolyte, allowing increased surface area for the electrodes. In this architecture, the Li^+ transport paths are shorter, reducing the electrical resistances across the battery [72]. Further, the ohmic drop of the interdigitated architecture is lower due to the smaller electrolyte/separator layer, leading to increased power.

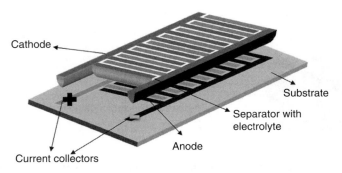

Figure 1.8 Schematic representation of the interdigitated battery architecture.

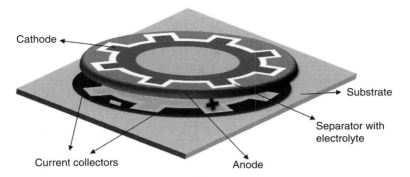

Figure 1.9 Schematic representation of the gear architecture.

Several works have been reported on interdigitated architectures fabricated by printing technologies, such as screen-printing, ink-jet printing and 3D-printing [73, 74].

Printed batteries based on interdigitated architectures have been fabricated based on lithium inks for anode, $Li_4Ti_5O_{12}$ (LTO), and cathode, $LiFePO_4$ (LFP), with high energy density, $9.7\,J\,cm^{-2}$, at a power density of $2.7\ mWcm^{-2}$ [74], values compatible with their use in microelectronics and biomedical devices.

The combination of printing technologies and batteries results in novel architectures, customizable for specific applications. In this sense, Figure 1.9 shows a "gear architecture", recently proposed, resulting from the application of the interdigitated architecture to circular batteries. This geometry is suitable for smartwatches, mobile phones and medical devices, among other applications [75].

Thus, printed batteries allow the development of novel architectures with optimized performance and better integration into specific devices [75].

1.4 Main Advantages and Disadvantages of Printed Batteries

1.4.1 Advantages

Printed batteries cannot compete with conventional batteries in applications where there are no size and shape limitations. On the other hand, they can fill the gap for small portable devices in which size, weight and improved integration in the device are some of the main requirements [35]. Other relevant areas for printed batteries are devices in which flexibility and stretch ability are required.

Other main motivations for the implementation of printed batteries is to reduce production costs and/or to achieve specific design features that printing technologies can provide [35].

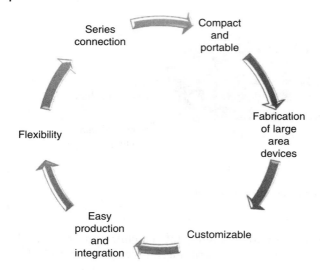

Figure 1.10 Main advantages of printed batteries.

The main advantages of printed batteries are summarized in Figure 1.10.

The main advantages of printed batteries are: easy production and integration; the fact that they are compact and portable, flexible and customizable; and the fact that they can be printed in series connection and allow fabrication of large-area devices (Figure 1.10). Printed batteries can be thinner than a millimeter, lighter than a gram, mechanically flexible and stretchable, allow specific designs and involve cost-effective production on a large scale [35].

One of the relevant advantages of printed batteries is the series connection of cells through printing technologies. In fact, the fabrication of printed batteries up to 15 cells has already been demonstrated [53].

1.4.2 Disadvantages

At the present moment, the higher cost of printed batteries in comparison to conventional batteries is one of the main drawbacks hindering the growth of their market share. Other problems are the need to develop new functional materials (inks) and to optimize processing.

It is expected that as production volumes increase and technology improves, the cost of printed batteries will decrease and they will compete with conventional batteries in price [76].

Figure 1.11 summarizes the current main disadvantages of printed batteries.

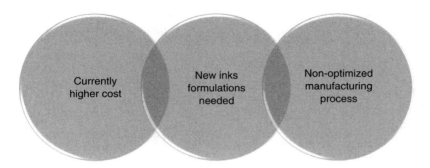

Figure 1.11 Main disadvantages of printed batteries.

In particular, the cost of printing batteries will be reduced through new manufacturing approaches (simplification of manufacturing methods) and through the development of lower-cost inks for the main components (electrodes and separator/electrolyte) [66].

1.5 Application Areas

There is an increasing number of powered small and portable devices such as RFID devices, microelectromechanical systems (MEMS), micro sensors, powered cards, smart toys and medical devices, among others; these are the main application areas of printed batteries (Figure 1.12).

Medical device applications of printed batteries include health-monitoring systems, wound-care and cosmetic uses, wireless patches for patient monitoring (electrocardiograms, monitoring of vital signs), and patient wristbands [77]. Being thin, flexible and disposable are common requirements of printed

Figure 1.12 Main application areas of printed batteries.

batteries for these applications. Further, it is necessary to adjust printed battery performance in terms of power and energy values, lifetime and discharge rate, depending on the application.

Thus, RFID tags require about 5.14 mW during the active state with a current consumption of 700 nA at 1.5 V, which represents a five-year operation for a 50 mAh battery [78]. In the case of transdermal-drug-delivery (TDD) systems printed batteries can power the integrated circuit that ensures the proper dosage control. For this system, the typical capacity value is 57 μA.cm^{-2} and, therefore, when using a printed battery with a capacity of 247 mWh, the system will continuously work for 12 days [35].

1.6   Commercial Printed Batteries

Printed batteries are already commercially available and, as previously indicated, are used in small portable electronic devices [35].

Different companies offer standard printed batteries and it is also possible to obtain customized batteries with different application requirements [35]. The current main producers of printed batteries are BrightVolt [79], Power Paper Ltd [80], Enfucell [55], Blue Spark [56], Imprint Energy [81] and Prelonic technologies [82].

The main commercially available printed batteries are non-rechargeable batteries based on zinc-manganese dioxide (Zn/MnO_2) with $ZnCl_2$ as an electrolyte. They are printed on plastic substrates and the open circuit voltage is typically over 1.6 volts for the batteries produced by Blue Spark and Enfucell [55, 56].

These batteries do not contain heavy metal components and the operation temperature range is −30 °C to 65 °C. BrightVolt produces high-energy-density printed batteries based on solid-state electrolyte.

All commercially available printed batteries are customizable in terms of voltage, size, shape, capacity and polarity and the shelf-life is from two to more than five years depending on the temperature [56].

Depending on its size and technology (Zn/MnO_2 and lithium), the price of printed batteries is in the range $2 to $5 for each battery.

1.7   Summary and Outlook

The growing interest in and applicability of printed batteries is related to the increasing interest in thin and flexible energy storage devices produced at low cost and based on eco-friendly materials in order to meet modern society's needs.

Printed batteries are an excellent alternative to conventional batteries for applications such as small and portable devices, radio frequency sensing, interactive packaging, medical devices and related consumer products, but it is necessary to further optimize the production cost and to develop new ink formulations to obtain higher capacities and voltages.

This chapter provided a general overview of the main characteristics, types, designs, advantages, disadvantages and applications of printed batteries, topics which will be further explored in the following chapters of this book.

There are both non-rechargeable and rechargeable printed batteries and the main types are lithium-ion batteries (LIBs) and $Zn-MnO_2$ batteries. One of the key issues for printed batteries is the development of suitable inks for high-performance printed battery components that are compatible with the desired print process, such as ink-jet or screen-printing. Thus, each ink should be prepared with optimized rheological properties for each printed technology, to ensure reliable flow, promote adhesion between the printed features, and provide the structural integrity needed for the functional battery.

The main challenges for printed batteries are the development of: high-performance ink formulations for each component, fully printed batteries, and novel battery architectures to improve device integration. Novel ink formulations for electrodes are continuously being reported in the literature but there is still a need to improve ink quality based on eco-friendly materials, to increase functional performance and to improve processability by different print technologies, including 3D printing. Further, it is essential to develop inks for the separator/electrolyte layer with strongly improved ionic conductivity, electrochemical stability and high mechanical strength.

The development of fully printed batteries is another very important challenge. In this case it is necessary to increase the compatibility of the inks for each component, leading to a more efficient printing process for the multilayer structure, a reduction of the internal impedance of the battery and an increase in cycling performance.

Finally, new design/architectures for printed batteries should be envisaged and fabricated with suitable delivered capacity and voltage for specific applications and improved device integration.

Despite these existing challenges, it can be concluded that printed batteries are part of current technological development, show a bright future for an increasing number of applications, and will enable a new generation of low-cost, portable and flexible applications.

Acknowledgements

This work was supported by the Portuguese Foundation for Science and Technology (FCT) in the framework of the Strategic Funding UID/FIS/04650/2013 project PTDC/CTM-ENE/5387/2014 and grants SFRH/BD/98219/2013 (J.O.) and SFRH/BPD/112547/2015 (C.M.C.). The authors thank the Basque Government Industry Department under the ELKARTEK Program for its financial support.

References

1 Osaka, T., Datta, M. (2000) *Energy Storage Systems in Electronics*, Taylor & Francis, London.

2 Chen, H., Cong, T.N., Yang, W., Tan, C., Li, Y., Ding, Y. (2009) Progress in electrical energy storage system: a critical review. *Progress in Natural Science* **19**, 291–312.

3 Luo, X., Wang, J., Dooner, M., Clarke, J. (2015) Overview of current development in electrical energy storage technologies and the application potential in power system operation. *Applied Energy* **137**, 511–536.

4 Toledo, O.M., Oliveira Filho, D., Diniz, A.S.A.C. (2010) Distributed photovoltaic generation and energy storage systems: a review. *Renewable and Sustainable Energy Reviews* **14**, 506–511.

5 Droege, P., Bocklisch, T. (2015) 9th International Renewable Energy Storage Conference. *IRES Hybrid Energy Storage Systems for Renewable Energy Applications, Energy Procedia* **73**, 103–111.

6 Dell, R., Rand, D.A.J., R.S.o. Chemistry (2001) *Understanding Batteries*, Royal Society of Chemistry, Cambridge.

7 Goodenough, J.B., Park, K.-S. (2013) The li-ion rechargeable battery: a perspective. *J. Am. Chem. Soc.* **135**, 1167–1176.

8 Wood III, D.L., Li, J., Daniel, C. (2015) Prospects for reducing the processing cost of lithium ion batteries. *J. Power Sources* **275**, 234–242.

9 Scrosati, B., Garche, J. (2010) Lithium batteries: Status, prospects and future, *J. Power Sources* **195**, 2419–2430.

10 Park, M., Zhang, X., Chung, M., Less, G.B., Sastry, A.M. (2010) A review of conduction phenomena in Li-ion batteries. *J. Power Sources* **195**, 7904–7929.

11 Costa, C.M., Silva, M.M., Lanceros-Mendez, S. (2013) Battery separators based on vinylidene fluoride (VDF) polymers and copolymers for lithium ion battery applications. *RSC Advances* **3(29)**, 11404–11417.

12 Gaikwad, A.M., Arias, A.C., Steingart, D.A. (2015) Recent progress on printed flexible batteries: mechanical challenges, printing technologies, and future prospects. *Energy Technology* **3**, 305–328.

13 Beidaghi, M., Gogotsi, Y. (2014) Capacitive energy storage in micro-scale devices: recent advances in design and fabrication of micro-supercapacitors. *Energy & Environmental Science*, **7**, 867–884.

14 Milroy, C., Manthiram, A. (2016) Printed microelectrodes for scalable, high-areal-capacity lithium-sulfur batteries. *Chemical Communications* **52**, 4282–4285.

15 Kamyshny, A., Magdassi, S. (2014) Conductive nanomaterials for printed electronics. *Small* **10**, 3515–3535.

16 Singh, M., Haverinen, H.M., Dhagat, P., Jabbour, G.E. (2010) Inkjet printing— process and its applications. *Advanced Materials* **22**, 673–685.

17 Sridhar, A., Blaudeck, T., Baumann, R.R. (2011) Inkjet printing as a key enabling technology for printed electronics. *Material Matters* **6**, 12–15.

18 Parc, printed and flexible electronics (2016) <https://www.parc.com/services/focus-area/flexible-and-LAE/>.

19 Cummins, G., Desmulliez, M.P.Y. (2012) Inkjet printing of conductive materials: a review. *Circuit World* **38**, 193–213.

20 Wunscher, S., Abbel, R., Perelaer, J., Schubert, U.S. (2014) Progress of alternative sintering approaches of inkjet-printed metal inks and their application for manufacturing of flexible electronic devices. *J. Mater. Chem. C* **2(48)**, 10232–10261.

21 Søndergaard, R.R., Hösel, M., Krebs, F.C. (2013) Roll-to-roll fabrication of large area functional organic materials. *J. Polymer Science Part B: Polymer Physics* **51(1)**, 16–34.

22 Choi, H.W., Zhou, T., Singh, M., Jabbour, G.E. (2015) Recent developments and directions in printed nanomaterials. *Nanoscale* **7**, 3338–3355.

23 Choi, J.-Y., Das, S., Theodore, N.D., Kim, I., Honsberg, C., Choi, H.W., Alford, T.L. (2015) Advances in 2D/3D printing of functional nanomaterials and their applications. *ECS J. Solid State Sci. Technol.* **4(4)**, P3001–P3009.

24 Khan, S., Lorenzelli, L., Dahiya, R.S. (2015) Technologies for printing sensors and electronics over large flexible substrates: a review. *IEEE Sens. J.* **15**, 3164–3185.

25 Ferrari, M., Ferrari, V., Guizzetti, M., Marioli, D. (2010) Piezoelectric low-curing-temperature ink for sensors and power harvesting. In P. Malcovati, A. Baschirotto, A. d'Amico, C. Natale (eds) *Sensors and Microsystems: AISEM 2009 Proceedings*, Springer, Dordrecht, Netherlands, 77–81.

26 Gonçalves, B.F., Costa, P., Oliveira, J., Ribeiro, S., Correia, V., Botelho, G., Lanceros-Mendez, S. (2016) Green solvent approach for printable large deformation thermoplastic elastomer based piezoresistive sensors and their suitability for biomedical applications. *J. Polymer Science Part B: Polymer Physics* **54**, 2092–2103.

27 SunJing-Bo, L.B., Xue-Guang, H., Kun-Peng, C., Ji, Z., Long-Tu, L. (2009) Direct write assembly of ceramic three dimensional structures based on photosensitive inks. *J. Inorganic Materials* **24**, 61147–1150.

28 Rose, A. (1981) Considerations in formulation and manufacturing of thick film inks. *Electrocomponent Science and Technology* **9**, 43–49.

29 Yang, C., Wong, C.P., Yuen, M.M.F. (2013) Printed electrically conductive composites: conductive filler designs and surface engineering. *J. Mater. Chem. C* **1(26)**, 4052–4069.

30 Buchdahl, R., Thimm, J.E. (1945) The relationship between the rheological properties and working properties of printing inks. *J. Applied Physics* **16**, 344–350.

31 Bauer, W., Nötzel, D. (2014) Rheological properties and stability of NMP based cathode slurries for lithium ion batteries, *Ceramics International* **40**, 4591–4598.

32 Shaw, D.J. (1966) *Introduction to Colloid and Surface Chemistry*, Butterworths, London.

33 Woo, K., Jang, D., Kim, Y., Moon, J. (2013) Relationship between printability and rheological behavior of ink-jet conductive inks. *Ceramics International* **39**, 7015–7021.

34 Beguin, F., Frackowiak, E. (2009) *Carbons for Electrochemical Energy Storage and Conversion Systems*, CRC Press, Boca Raton, FL.

35 Sousa, R.E., Costa, C.M., Lanceros-Méndez, S. (2015) Advances and future challenges in printed batteries. *ChemSusChem* **8**, 3539–3555.

36 Xie, K., Wei, B. (2014) Materials and structures for stretchable energy storage and conversion devices. *Adv. Mater.* **26**, 3592–3617.

37 Yan, C., Lee, P.S. (2014) Stretchable energy storage and conversion devices. *Small* **10**, 3443–3460.

38 Oudenhoven, J.F. M., Baggetto, L., Notten, P.H.L. (2011) All-solid-state lithium-ion microbatteries: a review of various three-dimensional concepts. *Adv. Energy Mater.* **1**(1), 10–33.

39 Lawes, S., Riese, A., Sun, Q., Cheng, N., Sun, X. (2015) Printing nanostructured carbon for energy storage and conversion applications. *Carbon* **92**, 150–176.

40 Sousa, R.E., Oliveira, J., Gören, A., Miranda, D., Silva, M.M., Hilliou, L., Costa, C.M., Lanceros-Mendez, S. (2016) High performance screen printable lithium-ion battery cathode ink based on C-LiFePO4. *Electrochimica Acta* **196**, 92–100.

41 Park, M.-S., Hyun, S.-H., Nam, S.-C. (2007) Mechanical and electrical properties of a LiCoO2 cathode prepared by screen-printing for a lithium-ion micro-battery. *Electrochimica Acta* **52**, 7895–7902.

42 Kang, K.-Y., Lee, Y.-G., Shin, D.O., Kim, J.-C., Kim, K.M. (2014) Performance improvements of pouch-type flexible thin-film lithium-ion batteries by modifying sequential screen-printing process. *Electrochimica Acta* **138**, 294–301.

43 Kim, H., Auyeung, R.C.Y., Piqué, A. (2007) Laser-printed thick-film electrodes for solid-state rechargeable Li-ion microbatteries. *J. Power Sources* **165**, 413–419.

44 Zhao, Y., Zhou, Q., Liu, L., Xu, J., Yan, M., Jiang, Z. (2006) A novel and facile route of ink-jet printing to thin film SnO2 anode for rechargeable lithium ion batteries. *Electrochimica Acta* **51**, 2639–2645.

45 Zhao, R., Liu, J., Gu, J. (2015) The effects of electrode thickness on the electrochemical and thermal characteristics of lithium ion battery. *Applied Energy* **139**, 220–229.

46 Singh, M., Kaiser, J., Hahn, H. (2015) Thick electrodes for high energy lithium ion batteries. *J. Electrochem. Soc.* **162**(7), A1196–A1201.

47 Orikasa, Y., Gogyo, Y., Yamashige, H., Katayama, M., Chen, K., Mori, T. *et al.* (2016) Ionic conduction in lithium ion battery composite electrode governs cross-sectional reaction distribution. *Scientific Reports* **6**.

48 Willert, A. (2009) Printable batteries. <http://www.fraunhofer.de/content/dam/zv/en/documents/rn7_FERTIG_tcm63–13052.pdf>.

49 Kil, E.-H., Choi, K.-H., Ha, H.-J., Xu, S., Rogers, J.A., Kim, M.R., *et al.* (2013) Imprintable, bendable, and shape-conformable polymer electrolytes for versatile-shaped lithium-ion batteries. *Advanced Mater.* **25**, 1395–1400.

50 Hassoun, J., Lee, D.-J., Sun, Y.-K., Scrosati, B. (2011) A lithium ion battery using nanostructured Sn–C anode, LiFePO4 cathode and polyethylene oxide-based electrolyte. *Solid State Ionics* **202(1)**, 36–39.

51 Gaikwad, A.M., Steingart, D.A., Nga Ng, T., Schwartz, D.E., Whiting, G.L. (2013) A flexible high potential printed battery for powering printed electronics. *Applied Physics Letters* **102(23)**, 233302.

52 Oakes, L., Hanken, T., Carter, R., Yates, W., Pint, C.L. (2015) Roll-to-roll nanomanufacturing of hybrid nanostructures for energy storage device design. *ACS Appl. Mater. Interfaces* **7**, 14201–14210.

53 Huebner, G., Krebs, M. (2015) Printed, flexible thin-film-batteries and other power storage devices. In S. Logothetidis (ed) *Handbook of Flexible Organic Electronics*, Woodhead Publishing, Oxford, 429–447.

54 Harrop, P., Zervos, H. (2009) *Batteries, Supercapacitors, Alternative Storage for Portable Devices 2009–2019*, IDTechEx.

55 Enfucell (2014) Enfucell Official Homepage. <http://www.enfucell.com/>).

56 Blue Spark (2014) Blue Spark Powering Innovation Corporation Official Homepage. <http://www.bluesparktechnologies.com>.

57 Scrosati, B., Abraham, K.M., van Schalkwijk, W.A., Hassoun, J. (2013) *Lithium Batteries: Advanced Technologies and Applications*, John Wiley & Sons, Hoboken, NJ.

58 Vincent, C.A. (2000) Lithium batteries: a 50-year perspective, 1959–2009. *Solid State Ionics* **134**, 159–167.

59 Winter, M., Brodd, R.J. (2004) What are batteries, fuel cells, and supercapacitors? *Chemical Reviews* **104**, 4245–4270.

60 Besenhard, J.O. (2008) *Handbook of Battery Materials*, Wiley-VCH, Weinheim.

61 Tam, W.G., Wainright, J.S. (2007) A microfabricated nickel–hydrogen battery using thick film printing techniques. *J. Power Sources* **165**, 1481–488.

62 Minakshi, M., Ionescu, M. (2010) Anodic behavior of zinc in Zn-MnO2 battery using ERDA technique. *International J. Hydrogen Energy* **35**, 7618–7622.

63 Braam, K., Subramanian, V. (2015) A stencil printed, high energy density silver oxide battery using a novel photopolymerizable poly(acrylic acid) separator. *Adv. Mater.* **27**, 689–694.

64 Berchmans, S., Bandodkar, A.J., Jia, W., Ramirez, J., Meng, Y.S., Wang, J. (2014) An epidermal alkaline rechargeable Ag-Zn printable tattoo battery for wearable electronics. *J. Mater. Chem. A* **2**, 15788–15795.

65 Yan, J. (2015) *Handbook of Clean Energy Systems*, 6 vols, John Wiley & Sons, Ltd, Chicester.

66 Kipphan, H. (2001) *Handbook of Print Media: Technologies and Production Methods*, Springer, Berlin.

67 Roberts, M., Johns, P., Owen, J. (2013) Micro-scaled three-dimensional architectures for battery applications. In Y. Abu-Lebdeh, Davidson, I. (eds) *Nanotechnology for Lithium-Ion Batteries*, Springer US, Boston, MA, 245–275.

68 Yan, C., Wang, X., Cui, M., Wang, J., Kang, W., Foo, C.Y. *et al.* (2014) Stretchable silver-zinc batteries based on embedded nanowire elastic conductors. *Adv. Energy Mater.* **4(5)**, 1301396– n/a.

69 Gaikwad, A.M., Zamarayeva, A.M., Rousseau, J., Chu, H., Derin, I., Steingart, D.A. (2012) Highly stretchable alkaline batteries based on an embedded conductive fabric. *Adv. Mater.* **24**, 5071–5076.

70 Hahn, R., Reichl, H. (1999) Batteries and power supplies for wearable and ubiquitous computing. *Digest of Papers. The Third International Symposium on Wearable Computers*, 168–169.

71 Pikul, J.H., Gang Zhang, H., Cho, J., Braun, P.V., King, W.P. (2013) High-power lithium ion microbatteries from interdigitated three-dimensional bicontinuous nanoporous electrodes. *Nat. Commun.* **4**, 1732.

72 Arthur, T.S., Bates, D.J., Cirigliano, N., Johnson, D.C., Malati, P., Mosby, J.M. *et al.* (2011) Three-dimensional electrodes and battery architectures. *MRS Bulletin* **36(07)**, 523–531.

73 Izumi, A., Sanada, M., Furuichi, K., Teraki, K., Matsuda, T., Hiramatsu, K. *et al.* (2014) Rapid charge and discharge property of high capacity lithium ion battery applying three-dimensionally patterned electrode. *J. Power Sources* **256**, 244–249.

74 Sun, K., Wei, T.-S., Ahn, B.Y., Seo, J.Y., Dillon, S.J., Lewis, J.A. (2013) 3D printing of interdigitated li-ion microbattery architectures. *Adv. Mater.* **25**, 4539–4543.

75 Miranda, D., Costa, C.M., Almeida, A.M., Lanceros-Méndez, S. (2016) Computer simulations of the influence of geometry in the performance of conventional and unconventional lithium-ion batteries. *Appl. Energy* **165**, 318–328.

76 He, X. (2016) *Flexible, Printed and Thin Film Batteries 2016–2026: Technologies, Markets, Players*, IdTechEx.

77 Wei, X., Liu, J. (2008) Power sources and electrical recharging strategies for implantable medical devices. *Frontiers of Energy and Power Engineering in China* **2(1)**, 1–13.

78 Namjun, C., Seong-Jun, S., Sunyoung, K., Shiho, K., Hoi-Jun, Y. (2005) A 5.1-/spl mu/W UHF RFID tag chip integrated with sensors for wireless environmental monitoring. In *Proceedings of the 31st European Solid-State Circuits Conference, 2005*, 279–282.

79 BrightVolt (2016) <www.brightvolt.com>.

80 Power Paper Ltd (2016) <http://powerpaper.cn/>.

81 Imprint Energy (2016) <www.imprintenergy.com/>.

82 Prelonic technologies (2016) <www.prelonic.com/>.

2

Printing Techniques for Batteries

Andreas Willert[1], Anh-Tuan Tran-Le[2], Kalyan Yoti Mitra[2], Maurice Clair[3], Carlos Miguel Costa[4,5], Senentxu Lanceros-Méndez[4,6] and Reinhard Baumann[1,2]

[1] Department of Printed Functionalities, Fraunhofer ENAS, Chemnitz, Germany
[2] Department of Digital Printing and Imaging Technology, Chemnitz University of Technology, Germany
[3] 3D-Micromac AG, Chemnitz, Germany
[4] Center of Physics, University of Minho, Gualtar campus, Braga, Portugal
[5] Center of Chemistry, University of Minho, Gualtar campus, Braga, Portugal
[6] BCMaterials, Basque Center for Materials, Applications and Nanostructures, Spain

2.1 Introduction/Abstract

The printing process involves the application of a specific ink onto a chosen substrate to form a predefined patterned layer of a certain layer thickness. After deposition of the ink the deposited ink needs to be dried, i.e. the solvents needed for the printing process itself need to be evaporated.

In Figure 2.1 numerous printing and patterning technologies that are used in production environments are shown. Not all of them can be used for printing batteries or the layers thereof. The left axis (Sheet-fed to Web-fed) indicates the application area. Sheet-based processes usually work on single sheets of material. Each single sheet needs to be aligned for the printing process. For increased speed, web-based material is commonly used because handling will become easier and therefore handling and printing speed can be increased.

All printing and patterning technologies were developed long ago, when printing of batteries was not even thought of. Therefore, optimization processes targeted goals other than technical application of functional layers. Nevertheless, developments to date are a reasonable basis for adopting printing technologies for the deposition of material layers that can, for example, be employed in the manufacture of batteries.

Printed Batteries: Materials, Technologies and Applications, First Edition.
Edited by Senentxu Lanceros-Méndez and Carlos Miguel Costa.
© 2018 John Wiley & Sons Ltd. Published 2018 by John Wiley & Sons Ltd.

Printing and patterning cloud

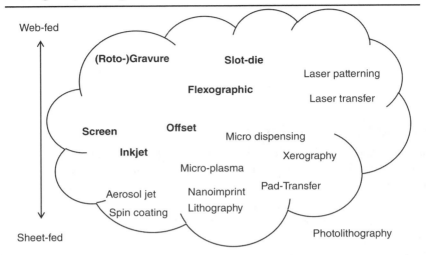

Figure 2.1 There are a number of established printing and patterning technologies; some can be employed for printing battery materials.

In this chapter, the most common printing techniques are addressed, while substrate and ink issues are discussed in detail in Chapter 3.

2.2 Materials and Substrates

The increasing use of primary and secondary batteries in a wide range of applications from smartphones and computers to electric vehicles leads to the need to improve their autonomy and to adjust their manufacturing process to specific application needs. Thus, additive manufacturing allows the fabrication of batteries for specific electronic devices [1, 2].

It is estimated that the global market in secondary batteries will grow more than 6% each year mainly due to the development of new materials but also due to the application of large-scale printing methods [3–5].

The junction between batteries and printing techniques results in printed batteries for areas such as radio-frequency identification (RFID) devices, trans-dermal-drug-delivery (TDD) systems, powered cards, smart toys and sensors, among others [6, 7].

The main types of printed batteries are lithium-ion, Zn/MnO_2 and Zn/Ag, which differ mainly in their constitutive materials [7].

Independently of the battery type, all printed batteries are composed of two different current collectors, two electrodes (anode and cathode) and the separator, as shown in Figure 2.2.

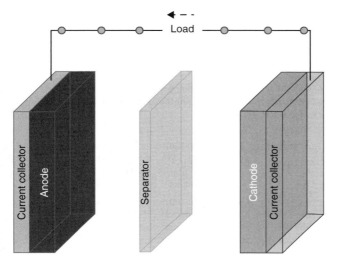

Figure 2.2 Schematic representation of the main components of a battery. (*See insert for color representation of the figure.*)

In the case of printed batteries, all materials should first be transformed into inks in order to be printable. Table 2.1 shows the main constituents of the different components of printed batteries.

Independently of the battery type, these batteries can be printed on different substrates such as plastic based on PET, PEN, polyimide, paper and fibrous substrates, depending on the device application [19, 20]. A sufficient sealing of the substrate, especially for lithium-based systems, is needed.

Ideally, these substrates should be flexible and chemically stable, as well as not suffering modifications at the temperatures needed for curing the inks [20, 21]. Further, substrates should be chosen taking into account the printing technique and device design [22]. Typically, it is necessary to apply a surface treatment to the different substrates in order to improve the adhesion of the ink and increase the quality of the printing [23].

Generally, the surface treatment is applied to the different substrates to reduce their hydrophobicity since the surface energy value determines the interaction between the ink and the substrate and defines the printing quality [23].

Polymer substrates are the most frequently used in printed batteries due to their high resistance to chemicals, and their light weight and flexibility [7].

2.3   Printing Techniques

All printing techniques that are employed for printing batteries were developed for applications in the graphic arts industry. The requirements of functional printing are fundamentally different to those of graphics. Graphic arts

Table 2.1 Main components and properties of different printed battery types.

Component	Constituent	Function/property	Examples	Refs.
Metallic current collector	Metal for negative and positive electrode	Conducts/transmits current from or to an electrode. Characterized by high electrical conductivity.	Most used: copper (negative) and aluminum (positive)	[8]
Electrodes	Anode active material	Absorption of ions.	Lithium-ion batteries: $Li_4Ti_5O_{12}$ and graphite. Other types: Zn or nickel	[9–11]
	Cathode active material	Availability of ions for the charging process and responsible for the cell capacity.	Lithium-ion batteries: $LiCoO_2$, $LiFePO_4$, among others. Other types: MnO_2 or Ag_2O	[12, 13]
	Conductive material	Increases the electrical conductivity of the electrode.	Conductive carbon black	[14]
	Polymer binder	Improves the mechanical strength and the adhesion of the electrode inks to the current collector.	PVDF, PTFE, PEO, PMMA	[15]
Separator	Polymer membrane	Avoids short-circuit.	Silica-based ionogels containing ionic liquid for lithium-ion type; PAA-based KOH gel electrolyte for Zn/MnO_2 type; 10 M KOH + ZnO for Zn/Ag type; $ZnCl_2$ for Zn/MnO_2 primary cell	[12, 16–18]
	Salts and/or electrolyte solution	Enables and increases the ionic conductivity value.		

addresses solely the human eye and its perception. Therefore, all printing technologies have been optimized for deposition of small dots of color that are perceived by the human eye as "nature-like". To achieve this goal printing technologies deposit layers of materials. These deposition properties as well as the patterned manner of the deposition of most printing technologies are the basic properties that make them so suitable for the well-controlled deposition of material. We call this approach "additive", indicating that we just deposit material at the pattern geometry where it is really needed. In contrast "subtractive" lithographic processes first cover the whole area with a material film, pattern this film, remove some parts of it and carry out a chemical etching process to unveil areas.

All printing technologies have unique features in respect to feature size, accuracy, layer thickness and printing speed, and also the capability to handle inks with a certain particle size.

2.3.1 Screen Printing

Originally, the screen for the printing was made out of silk; therefore, often this printing technique is called "silk screen printing". The advantage of this printing technology is that it can cope with rough and big particles (up to several tens of µm) [24]. Furthermore, the deposited layers may vary from less than 1 µm to more than 100 µm – in just one printing step [25]. The inks that can be processed are in a viscosity range from 100 cP to 100,000 cP [26]. Some applications can be found in [27–30].

Screen printing is used in graphic arts for the printing of textile fabrics, advertisement panels or even glass windows. Materials can be flat (plastics; fiber-based, such as paper or fabrics; or metal) or 3D-shaped (such as bottles, encasings or even devices). In these applications much material is needed to colorize the textile fabrics and to ensure a long-lasting illustration. In all applications the robustness of the screen printing technique is the advantage. Also the screen is scalable to sizes up to several square meters.

It originated with stencil printing (see below). Especially for textile fabrics, the technique developed to include a supporting grid for the stencil layer to also cover isolated areas. In this development silk was the supporting material for the stenciled areas.

Nowadays the silk is replaced by polymeric or metal meshes.

Layer thicknesses of screen printing can be in the range 20 nm to 1 mm [31].

There are more than 50 known parameters that influence the screen printing process [32]. The most important parameters are given in Table 2.2.

Basically there are two main setups for screen printing: flatbed and rotary screen printing. The common features and the differences are discussed below. There are also other setups established in the industry, such as employing a rotary screen for flatbed printing. Also, first approaches are shown to use screen printing for material deposition onto 3D-shaped objects.

Screen printing has been used in printing layers of batteries on various kinds of substrate: plastics [34, 35], aluminum [36], paper [37], glass [15], textile [38].

2.3.1.1 Flatbed

In Figure 2.3 the setup of the screen printing process is shown.

Screen printing is a technology in which the ink is transferred from one side of the printing form onto the substrate that is located beneath the other side of the printing form.

In flatbed screen printing the substrate is fixed to a base plate so that the printing area is flat. With a distance in the order of millimeter the screen is held in parallel (jump distance). The screen material is fixed in a metal frame

Table 2.2 Selection of important parameters for screen printing [33].

Origin	Parameter
Screen printing machine	Printing speed
	Snap-off distance
	Repeatability accuracy
Substrate	Material
	Surface properties (pre-treatment)
Printing ink	Viscosity
	Homogeneity
	Composition
	Absorption characteristics
	Evaporation properties of solvents
	Drying properties
Screen	Material
	Tension
	Elasticity
	Threat density
	Treat thickness
	Mesh width
	Thickness of stencil material
	Jump height
Squeegee	Hardness
	Alignment angle
	Pressure

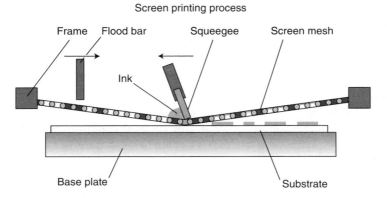

Figure 2.3 Principle of the flatbed screen printing process.

to enable a defined tension for the screen mesh. The screen itself consists of a fabric of threads enabling the transfer of material through open meshes. To determine the pattern of the transferred material one side of the meshes is coved by a screen emulsion. Therefore, the second function of the screen is to mount the stencil elements. On top of the screen the ink is placed and moved by a squeegee. The squeegee applies a define pressure onto the screen, so it gets in direct contact to the substrate in the printing line. Outside this line there is a gap between them, supporting the ink transfer process. The transferred ink will remain on the substrate if the adhesion and cohesion forces are well balanced so that only a minor fraction will remain on the screen.

The ink will be at the left side of the screen after the printing step. To move it back to the right side a flood bar is used (it is mounted on the left-hand side of the squeegee; in moving right to left it is lifted; only for pushing the ink it is lowered onto the screen mesh) moving in the opposite direction to the squeegee. In this process the material is distributed over the whole mesh and the open mesh areas are filled with ink. In the printing process the stop position of the ink can be optimized: either kept to the left side to restrict evaporation of the solvent or flooded over the printing area. In the latter case the evaporation effects are much stronger.

2.3.1.2 Rotary

This basic principle has been modified for industrial processes to gain higher throughput. For example, for web-fed applications the screen mesh is not flat but built as a cylinder, the so-called rotary screen. The advantage is that after each rotation the printing can be continued. This is, for example, used for the screen printing of textiles in lengths of several hundred meters without any disruption. A schematic drawing is given in Figure 2.4.

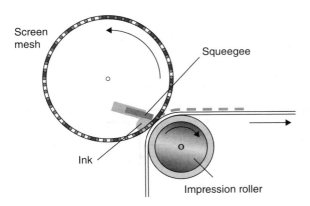

Figure 2.4 Principle of a rotary screen printing process.

The relative movement of squeegee and substrate web is still the same in this setup; the squeegee is fixed but the screen mesh as well as the substrate are being moved. Therefore, the transfer across the screen mesh is technologically identical.

2.3.1.3 Screen Mesh

One fundamental parameter in the screen printing process is the screen itself. The amount of ink is determined by the setup of the screen as well as the maximum particle size that can be transferred.

The screen type is usually referred to as a number like

$$120 - 24 \text{ YPW}$$

In this expression 120 is the threads per cm (in the U.S. per inch) and 24 is the diameter of the threads in μm. The letter "Y" indicates a yellow mesh material while a "W" would indicate a white one. "PW" is indicating a plain weave while alternatively "TW" stands for twill weave. These parameters are shown in Figure 2.5.

The next parameter of the screen printing form is the emulsion layer, which covers all the mesh areas that should not transmit any material and therefore determines the print pattern. In Figure 2.6 this layer that prevents any ink transfer at these locations is shown.

To increase the transferred ink volume it is common to thicken this emulsion layer to more than the typical 8 μm [39].

When calculating the transferred ink volume one might take into account that only about 70% of the calculated mesh volume is emptied during the printing process [39]. Taking this into calculation, [39] presents the formula for calculation as

$$h_w = 1.3\, h_s\, k_r \tag{1}$$

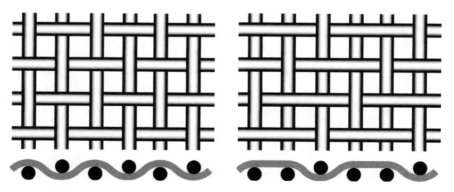

Figure 2.5 Left – plain (PW) and right – twill (TW) 1:2 weave: black: side view of thread.

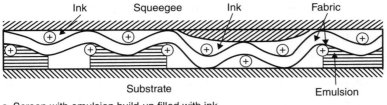

Ink   Squeegee   Ink   Fabric

Substrate   Emulsion

a. Screen with emulsion build-up filled with ink

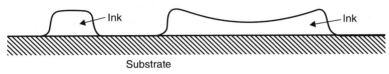

Ink   Ink

Substrate

b. Ink deposit on substrate

Figure 2.6 Screen properties in transferring ink [39].

where h_s is the fabric thickness (usually $2 \times d_w$); d_w is the diameter of wire in fabric

k_r is the relative fabric openness; $k_r = [(m - d_w)/m]^2$

with $m = 1/M$ is the wire pitch of fabric

M is the mesh count per unit length

Equation (1) is usually simplified to $h_w = h_s\, k_r$ [39–41].

2.3.1.4 Squeegee

The squeegee is the most important parameter besides the screen mesh and ink when the printing process is performed. The squeegee has several tasks to fulfill:

a) The squeegee moves the ink across the screen, fills the open meshes and moves inks from the closed meshes.
b) By applying pressure onto the screen the squeegee generates a contact line between ink, screen and substrate for ink transfer.
c) The speed and properties of the squeegee influence the ink transfer.

Depending on the pattern feature size a low printing speed is recommended for a high-resolution pattern. In battery printing most patterns are very coarse so that a higher squeegee speed can increase the transferred layer thickness.

There are two main important properties of the squeegee: hardness and shape [42]. To easily differentiate the hardness there are, for example, red (60°–65° shore), green (70°–75° shore), and blue (75°–80° shore) elastomer materials used. The higher the shore number the harder (i.e. less elastic) the elastomer is. A very high number ensures very high pressure with low deformation of the squeegee in the printing process, resulting in a very high feature size [43].

With increasing substrate hardness, the squeegee material should be softer to enable a reliable ink transfer.

Employing a rectangular squeegee profile will result in a narrow contact line while a rounded profile will have a broader contact line enabling a higher volume of ink transfer. Therefore, the type of squeegee has a big influence on the transferred layer thickness [44].

Also important for the printing process is the angle of the squeegee with respect to the screen [45]. The variation of this parameter will also influence the resulting pressure within the ink during the printing process and consequently the amount of ink transferred.

2.3.2 Stencil Printing

Stencil printing is shown in Figure 2.7.

In a stencil pattern there are open and covered areas on the substrate. Ink is transferred through every uncovered area. The big disadvantage compared to screen printing is that due to the lack of any mounting material all stencil areas need to be connected to each other. Therefore, it is impossible to have isolated covered stencil areas without at least a small connection to the other stencil material.

In stencil printing there are two options in applying the ink material. The first is spray coating onto the substrate. The covered areas are also inked and the ink stays on those areas. If there is a surplus of material it will flow towards

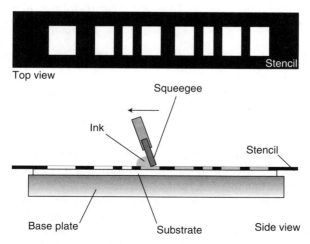

Figure 2.7 Schematic of stencil printing. The top view is given in the upper part of the sketch.

the non-covered areas and lead to an unpredictable layer thickness distribution. Furthermore, surplus material might also be swallowed between stencil and substrate resulting in spoilage printouts.

The second method of stencil printing is the use of a doctor blade. In this case the ink is deposited on one side of the stencil and moved by the doctor blade across the stencil form, filling the open areas. This approach is much more reliable compared to spray-coating application. Nevertheless, due to the abovementioned restrictions screen printing is the preferred method of material application.

2.3.3 Flexographic Printing

Letterpress printing (a technique of relief printing) is the oldest printing technology. Starting in the middle of fifteenth century, it was most commonly used in the book printing process. For the following centuries, letterpress printing was the prevalent printing technology for newspapers and printed documents for government, businesses and churches. In the early 1900s, a modified letterpress printing method called flexography became predominant in the printing industry, since it can be used to print on almost any type of substrate [30, 46–49].

2.3.3.1 Letterpress Printing
Letterpress printing is a mechanical process in which the highly viscous pasty ink is transferred onto the substrate via the hard printing elements (for example, lead lettering) using high pressure. In Figure 2.8, the ink is transferred from

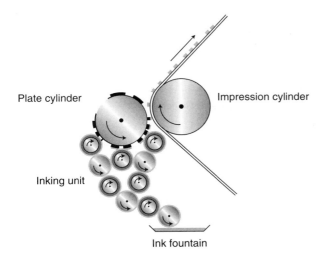

Figure 2.8 The principle of the letterpress printing process.

the ink container/fountain to the raised printing elements on the hard printing plate via inking rollers to homogenize the layer thickness. The impression roller consequently presses the substrate onto the printing plate. Finally, the ink is transferred/printed onto the substrate.

Based on the different pressing methods involved, letterpress printing presses and machinery are divided into hand presses and platen, high speed presses and web presses. There is also an indirect letterpress printing method (letterset), in which the ink is transferred from the printing plate to the substrate via a blanket cylinder.

2.3.3.2 Flexography

Flexography is a modified method of letterpress printing technology. It uses low viscosity inks and resilient, flexible printing plates and applies low pressure between plate cylinder and substrate. Due to the implementation of flexible printing plates, which are made from rubber and photopolymeric plastic, flexography can be used for printing on all types of paper, cardboard, rough surface packaging materials and fabrics.

The printout quality of flexography for graphic arts is lower compared to the products of offset printing. However, the development of new types of printing plate, adapted ink and technical printing press has remarkably improved the printout quality offered by flexography.

The flexography printing process shown in Figure 2.9 utilizes an anilox roller to transfer the ink from the ink container with a chambered doctor blade onto the soft printing plate. The print image is then transferred to the substrate by pressing the impression roller to the printing plate.

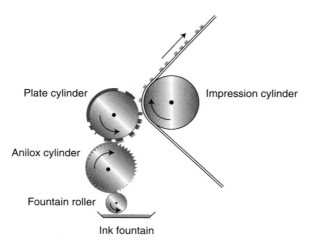

Figure 2.9 The principle of the flexography printing process.

The flexography requires a slight contact pressure to transfer ink from printing plate to substrate. The resilient and flexible printing plate provides a good printout result even with low contact pressure. It must be considered that deformation and wear of the flexible printing plate can lead to considerable dot-size increase and dot gain. Furthermore, the polymeric printing form may be damaged if incompatible solvents are used.

The rubber printing plates are manufactured by molding a matrix using natural rubber. The best quality rubber plate is produced by applying a laser to engrave the fully coated elastomer plate cylinder.

Photopolymer printing plates are manufactured using the UV light exposure process. The exposed photopolymer area is crosslinked and then washed-off. Single layer plates are made in thicknesses from 0.76 mm up to 6.35 mm depending on the print products. There is also a multilayer plate with the combination of a relatively hard thin layer plate and a compressible substructure. The multilayer plate is used for high-quality halftone printing.

Further Development of Letterpress/Flexography Printing

Traditional letterpress printing has obviously become insignificant due to the advantages of flexography printing (low viscous inks, many types of substrate, etc.). Many developments and enhancements have been achieved in the field of flexography printing. The quality of printed products in flexography printing has improved and is close to the high quality of offset and gravure printing. Very good progress has been made in the prepress area of flexography printing as a result of computer-to-plate technology. The application of sleeves as printing master also contributes greatly to good printout results.

2.3.4 Gravure Printing

Gravure printing started in the early fifteenth century and is a very old printing process. It is a very good printing technology for illustrations, producing very high image quality printouts. Typical products of gravure printing include magazines, catalogs, carrier bags, stamps, bank notes and security paper. Due to its high accuracy it is also widely employed in functional printing processes [50–52].

Contrary to letterpress printing technology, gravure printing technology has the distinctive feature that the image elements are etched/engraved onto the surface of the printing cylinder. These etched/engraved elements are called cells and the non-printing areas are at the original level. The first step of the printing process (Figure 2.10) is the inking of the entire gravure cylinder (both the printing and non-printing areas). The ink is then wiped out of the non-printing areas using a doctor blade. After the doctor-blading process, the ink just remains inside the cells of the gravure cylinder. Finally, the impression

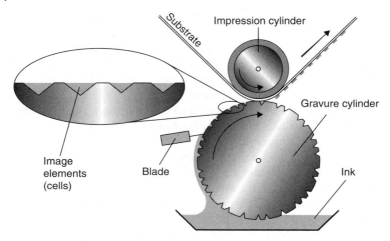

Figure 2.10 The principle of the gravure printing process.

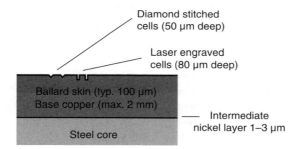

Figure 2.11 Cross cut of a gravure cylinder.

roller presses the substrate onto the gravure cylinder and the ink is transferred to the substrate by a high printing pressure and the adhesive force between the ink and the substrate keeps it there.

In Figure 2.10 the typical structure of a gravure cylinder is shown. The steel cylinder is coated with a very thin nickel layer (1–3 μm) and an electroplated base copper layer (2 mm). On the top of the base copper layer, there is one coating layer (100 μm) of engraving copper or Ballard skin layer which is engravable (cuttable). On the outmost side of the cylinder is the thin chromium protecting layer, which ensures a long life for the gravure cylinder (Figure 2.11).

The image to be printed is broken up into single printing elements, so-called printing cells. These cells are engraved into the gravure cylinder. The engraving process can be carried out by etching, electromechanical engraving or the laser engraving technique.

The advantages of gravure printing technology are the simplicity of the printing process, the high-quality printouts and the high output and consistency of products. However, a significant disadvantage is the expensive and time-consuming preparation process of the gravure cylinder. Therefore, progressive attempts have been made to replace the rigid gravure cylinder with quickly changeable printing plates or with sleeve technology. Laser engraving was also developed to provide a fast, high-precision engraving technique in gravure cylinder manufacturing.

2.3.5 Lithographic/Offset Printing

Lithographic printing is widely used for newspaper and job printing in print houses for moderate numbers of copies (1,000–300,000). It is a high-throughput technique, which is depicted in Figure 2.12.

The basic printing principle is to differentiate two areas on the printing plate as between their surface properties: an oleophilic and an oleophobic pattern. For the printing process, first, water is applied to the printing plate, covering all oleophobic areas. After this process the oil-based ink is applied to the printing plate, covering all oleophilic areas, while being repelled from all water-covered oleophobic areas. The inked printing plate is transferred onto a blanket cylinder and then applied to the substrate. A couple of rollers are applied to the water as well as to the ink supply.

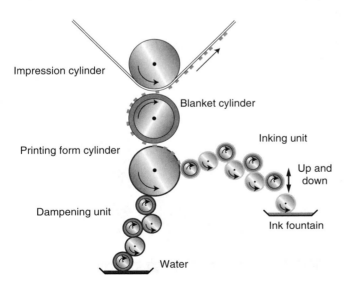

Figure 2.12 Lithographic/offset printing setup.

The whole process will generate some maculation copies until the equilibrium status in the printing process has been reached. This number might be some single copies up to several hundreds.

The main working principle is to use ink and supporting agent (for example, water) with oleophilic and oleophobic properties. These requirements generate an obstacle for battery production, so that no application is known.

2.3.6 Coating

A coating technique consists in applying a solution composed of different components, such as polymers and particles, within a polymer solvent, in which a film is obtained after solvent evaporation [53]. This technique can be divided into doctor blade (or tape casting) and spray coating, where the main differences between the techniques are the application procedure and the different ink properties, such as ink viscosity. For spray coating, the solution is sprayed onto the substrate by spray guns [54] and for doctor blade coating, the solution is spread through the blade by unidirectional movement on a substrate [55, 56]. Figure 2.13 shows a schematic representation of both techniques.

For both techniques, the formation of the film and its quality are influenced by the solution formulation, i.e. solvent type and concentration, solids loading and their dispersion, which affect the viscosity value, but also solvent evaporation temperature and surface energy of the substrate [57]. These technologies are also widely used for the manufacturing of organic solar cells [48, 58].

In the doctor blade technique, the coating speed is related to the shear stress and is an important parameter with linear speeds between 1 and $100\,mm.s^{-1}$ being typically needed for meniscus formation between blade and solution.

(a)　　　　　　　　　　　　　(b)

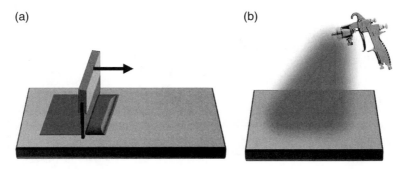

Figure 2.13 Schematic illustration of: a) doctor blade and b) spray-coating techniques.

The final thickness of the coating after solvent evaporation can be calculated through the following equation [59]:

$$d = \frac{1}{2}\left(g\frac{c}{\rho}\right) \qquad (1)$$

where g is the gap distance between the blade and the substrate, c is the concentration of the solid material in the ink in $g.cm^{-3}$ and ρ is the density of the material in the final coating in $g.cm^{-3}$. A decrease or increase in the blade gap can be correlated linearly with the thickness of the coating, which is also affected by the casting speed [60]. With this technique, it is possible to obtain thin film with a thickness of $5\,\mu m$ but the typical thickness of coatings is above $25\,\mu m$.

Spray coating is a very fast and simple technique where the process control parameters are air pressure, solution viscosity and evaporation rate of the solvent, as well as properties of the gun such as its geometry. Some authors state that the viscosity and surface tension of the solution are not particularly relevant issues for this technique since the air pressure is the most important factor and can be adjusted for each gun and solution [54].

Finally, the efficiency of the spray coating is determined by other parameters such as the distance between the nozzle and the substrate, and the flow rate of the solution [61].

Table 2.3 shows the most relevant advantages and disadvantages of the doctor blade and spray-coating techniques.

Taking into consideration the advantages and disadvantages presented in Table 2.3, both techniques are already used for printed batteries, where the main advantages common to both are their simplicity and facility for producing thin films in large areas [7].

Table 2.3 Main advantages and disadvantages of the doctor blade and spray-coating techniques.

Technique	Advantages	Disadvantages
Doctor blade [58, 59]	• Large area and good uniformity, • One pass, • Large viscosity range, • No/little waste of material.	• Without pattern, • High thickness > 1 μm.
Spray coating	• Large area, • Adjustable thickness, • Nearly independent of the substrate.	• Multiple passes, • Inhomogeneity of the film.

Both techniques have been applied for the development of printed batteries, in which the solvent concentration of the inks should be between 1 and $8\,mL.g^{-1}$, depending on the particle size, size distribution and polymer type [62].

2.3.7 Inkjet

There are three sub-sections in this section:

2.3.7.1 Inkjet Printing Technology and Applications
2.3.7.2 Selective View of the Market for Inkjet Technology
2.3.7.3 Advanced Applications: Printed Functionalities and Electronics

Keywords

Drop-on-Demand (DoD), Continuous Inkjet Technology (CIJ), Thermal Inkjet Technology (TIJ), Piezo-electric Inkjet Technology (PIJ), Super-fine Inkjet Technology (SIJ), roll-to-roll (R2R), sheet-to-sheet (S2S) and roll-to-sheet (R2S).

2.3.7.1 Inkjet Printing Technology and Applications

For the last two decades the most flexible of all the printing techniques has been inkjet printing technology. Inkjet printing technology has been categorized as a digital manufacturing technology that has several advantages over conventional printing technologies, such as the ability to print variable data on commercial articles such as bar codes, serial numbers, identification graphics, anti-counterfeiting elements, etc. on packaging. The technology also offers the possibility of producing documents which do not require high quantities of consumables (e.g. ink and substrate), and each document can be individualized as in small office home office (SoHo) applications. In this case, documents can be of any kind, ranging from printing of certificates, flyers, event bookings and even official documentation. Another advantage is the low requirement for ink and substrate in the context of small and home office as the target application areas. The next promising aspect of inkjet printing comes from the technological background itself – inkjet is a contactless printing technology, which means that there is no direct contact between the printhead and the substrate for the ink-transferring mechanism [25, 63–66]. Inkjet printing technology offers several drop volumes in the range pL to µL, printing resolutions ranging from 100 dpi to 5000 dpi, and resulting wet-layer thickness ranging from 50nm to 10µm depending on the ink used and the substrate implemented. These two basic properties of inkjet (technologically flexible and contactless) open up several possibilities for contactless printing over irregular or asymmetric shapes and forms, e.g. commercial objects such as bottles, edible products, irregular packages, etc. Besides this, while the technology offers contactless ink deposition, it can also have high deposition accuracy in the micrometer range ($10 \pm 5\,\mu m$). Also, due to the fact that the inkjet printing technology is implementation-wise very flexible and yet affordable, it is also widely used for research and development purposes in the state-of-the-art application field of "printed functionalities" and "printed electronics or microelectronics". Around the world, many researchers are focused on developing printed electronics products: active and passive devices such as capacitors, thin-film-transistors, resistors, inductors, antennae, etc.; sensors and detectors based on elemental factors such as humidity, pressure, stress, strain, light, gas concentration, thermal, etc.; energy harvesting/harnessing and power devices such as batteries, organic photovoltaics, piezo-electric actuators, etc.; and at a very complex level, even circuits such as device arrays and voltage-operated switching circuits based on logic gates. And on the other hand, development of printed functional layers such as printed sieves, lenses, haptics, photonics, etc. are other fascinating application areas.

In theory, there are different kinds of inkjet printing systems based on the backend technology, the drop-ejection/printing mechanism, the printhead system architectures and the application areas. Two of the well-known, typical kinds of inkjet printing technology are Drop-on-Demand "DoD" and Continuous Inkjet "CIJ".

CIJ works on the principle of continuous drop generation. It is triggered by a voltage-operated electronic transducer or actuator creating continuous impulse waves to force the ink out of the nozzle orifice in the form of a stream or jet. This stream or jet contains perturbation which breaks down in flight to drops of similar shape and volume. Furthermore, the drops become separated from each other with a specified distance and with certain drop velocity. They then enter the first electric field, where the drops are charged electrically with a single entity and now the charged drops enter a second electric field (processed with deflection) possessing the opposite charge entity. Depending on the intended digitalized image (whether it is to contain a non-printable or printable pixel), the opposite charge entity would not be activated (non-printable pixel) or activated (printable pixel). This process is also called deflection, and it allows the drops to be discriminated for printing from the continuous stream of flying drops. The drops that are not deflected go straight to the gutter, where they are redirected to the recirculation system and finally to the ink reservoir.

On the contrary, DoD inkjet technology works on a relatively simpler and very different principle. Discrimination of drops to be chosen to print a digitalized pixel is determined by the movement of the transducer or actuator itself. DoD-based inkjet systems can be classified into two further technologies based on the working mechanism – thermal inkjet technology (TIJ) or piezo-electric inkjet technology (PIJ). The schematics for such DoD- and CIJ-based inkjet systems can be seen in Figure 2.14 [25, 63, 65, 66].

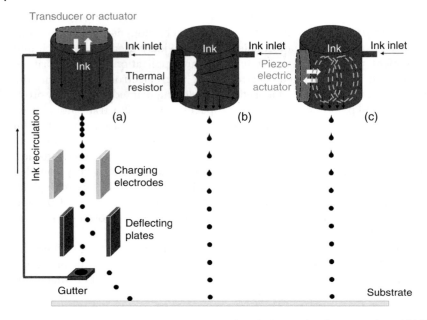

Figure 2.14 Scheme showing the basic principles of (a) CIJ technology, DoD-based (b) TIJ and (c) PIJ technology. (*See insert for color representation of the figure.*)

In the case of a thermal inkjet system, which is also known as bubble inkjet, the transducer or actuator is basically a voltage-operated thermal resistor (thermistor) which can be heated up to a very high temperature. The ink generally has to be in contact with this thermistor, thereby producing a bubble-like vacant structure at the interface of the ink and thermistor. This bubble, at this high temperature, expands and explodes in volume, and to compensate for the internal volume expansion and the pressure within the ink reservoir, the ink drops are effectively forced out of the nozzle. The amount of increase in the temperature of the thermistors is dependent on the application of voltage to them, which is controlled by the application of electronic signals for either a printable pixel digitalized image or non-printable pixel digitalized image. Hence, it can be said that the drop-ejection system based on thermal inkjet technology is very complicated and is limited in relation to the choice of materials. Only certain solvent-based inks compatible with this technology are especially designed for the printing process. On the positive side, inkjet printers are very suitable for low-quantity SoHo applications and the initial investment or replacement costs are relatively low.

The mechanism of the drop ejection in the case of piezo-electric-based DoD inkjet system is very simple and can be considered similar to thermal inkjet technology. The main principle behind drop ejection is the same, that is, compensation for the expanded ink volume within the ink reservoir. As already indicated here the ink movement is controlled by the impulse pressure waves triggered by the piezo-electric transducer or actuator. When voltage is applied to such piezo-electric actuators in positive and negative entity, their neutral shape starts to deform inwards or outwards. This inwards movement retracts the ink from the main ink chamber and the outwards movement creates a pressure wave impulse which propagates through the stagnant ink directed to the walls of the ink chamber. Then, as the pressure wave retreats it is opposed by the second phase of the pressure wave impulse propagated from the piezo-electric actuator. When these different-phased pressure wave impulses meet, they superimpose over each other. Thus, there is an increase in the ink volume inside the chamber and in order to compensate for this, increment drops are jetted out of the nozzle orifice. These drops jetted out of the nozzle place themselves at equal distances from each other with a constant drop velocity, finally reaching the target substrate. This movement of the actuator is controlled by electronic signals that are generated by the digitalized image containing either printable or non-printable pixels (image or non-image area). Piezo-electric-based DoD inkjet systems are generally more user friendly and there is an enormous choice of ink materials that can be used for the printing process whether for small SoHo applications or large industrial applications.

Another kind of inkjet technology which is very different from either TIJ- or PIJ-based actuation technology is recently grabbing attention. This is called Super-fine Inkjet Technology (SIJ), which is based on the principle of electro-hydrodynamic inkjet technology. In this technology, the basic principle is to control the droplet and its size with the help of electrostatic forces and without any involvement of mechanical action for drop generation or ejection. Such a scheme is shown in Figure 2.15 [67–71].

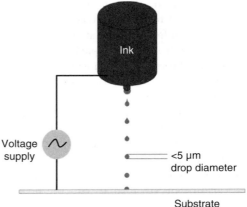

Figure 2.15 Scheme showing the basic principle of SIJ based on electro-hydrodynamic inkjet technology.

The main motivation of SIJ is to extend the limit of printing resolution for inkjet printing technology. SIJ creates an oscillating electric field or electro-static forces between the substrate plate and the ink reservoir (ink meniscus within the nozzle), thereby charging the ink and controlling the ink meniscus to form drops by altering the electric field or the charge accumulation. The voltage actuation can be oscillated and this influences the strength of the electric field, resulting in a definite ink drop size. The electrical parameters of the substrate material and its conductivity will also affect the size of the ejected droplet. When the substrate material is insulating or non-conductive, e.g. plastic, then the charge accumulation at the substrate surface also occurs during the process of printing. For SIJ also, the waveform determines the size, volume and shape of the drop. In the case of a unipolar waveform, the charge accumulation will decrease the effective electric field at the ink meniscus leading to a smaller droplet size. Some of the very general parameters which can be varied for altering the drop size and shape are as follows: voltage (value and amplitude) for the oscillating electric field, waveform (sinusoidal or square depending on the ink), frequency of oscillating electric field, printing speed and nozzle to substrate distance. SIJ technology operates over a wide range of fluidic viscosity 1–30 mPas. The drop volume can be varied between 0.1–1 fL with a small nozzle tip < 10 μm and printing speed of ∼1 mm/s. With this technology, printed line width of < 5 μm is easily achievable, with a drop size of 1–5 μm. And in some cases, sub-μm line widths are achieved when the drop size is further reduced, with drops exhibiting no spilling/splashing effect once they land on the substrate. SIJ is another very promising technology, which is presently used intensively from research laboratory scale to industrial scale [67–71].

Inkjet can no longer be considered as suitable only for sheet-to-sheet (S2S) printing technology in DIN-A4-size format. It has grown massively during the last 20 years and has proven its existence as a relevant, mature industrial technology. The technology has been upscaled to wide-format printing platforms where rolls of textiles, posters, banners, decorative table cloths, wallpapers, etc. are manufactured on an industrial level. Printing width has increased to > 5 m and production capacity to > 120 m^2/h. Moreover, the basic principle has stayed the same, with most of the printer manufacturing companies developing their printing systems based on the DoD inkjet system mounted with a robotic arm which can move along the printing axis, and this is combined with the curing process. Mostly, ultraviolet and thermally curable inks are used in various commercial applications. There is also a huge choice of compatible inks with specific fluidic properties which generally fit the inkjet printing category. The print and manufacturing process for certain special applications can be modified according to customer needs. These kinds of printing systems can also be in S2S format with the post-printing processes such as slitting, folding and perfecting, etc. already integrated into the system.

Furthermore, there are various other inkjet systems based on DoD and CIJ which are very much based on industrial platforms, and production in this case is much higher. The main focus in this kind of printing system is to print fast, with the unwanted compromise on printing quality. Some common examples would be printed products such as newspapers, low-budget magazines, books, pharmaceutical items, automobile mechanical parts, labels, bar and QR codes, gloss or varnish, and printing on irregular articles such as bottles, chocolates, eggs, etc. In all these production applications the requirement of fast printing speeds is inevitable and strictly required. Therefore, in many situations there is a need to develop R2R or R2R-like inkjet print production lines. The production line is custom designed according to the product fabrication requirements. The printing resolution, speed and print width of the printing process can be defined by the printhead type – number of nozzles, arrangement of nozzles and drop volume, number of printheads stacked beside each other, native resolution of the printhead or angular rotation of the printhead setup, skewing of the digital image, and droplet-jetting frequency with respect to the R2R moving web. Once the inkjet-printed layer is deposited on the moving web, it is essential that the ink is dried using a post-printing curing method installed inline after single- or multi-color printing setup. The amount or duration of post-printing curing is always defined by the number of printing steps. Also, for such printing setups, the main motivation is to vary the digital data or content for the process, e.g. in the case of labeling and code development by printing. These applications come under the category of variable data printing. The process can even be adjusted for printing over irregularly shaped items, along with variations in the data deposited on the surface of the article. In some extreme cases, such R2R inkjet printing systems are also integrated with an R2R conventional process, e.g. gravure, flexography or offset printing press. The speed of the conventional R2R printing system is incomparably fast, and to complete the printing value chain, variable data must be integrated into the product. In this scenario, additional printhead bars (inkjet printing unit) are added at the end of the print workflow so that all the variable or special data can be deposited onto the products at the final phase. This kind of modification is very logical and yet impressive for high-production pilot lines.

Finally, a special application of inkjet printing technology is meeting customized general public needs, e.g. low-budget commercial products that can be built by rapid prototyping or even by three-dimensional (3D) printing. The product range is basically enormous, e.g. figures, medical and surgical applications, tools and machinery, sports applications, aerospace applications, product conceptualization, architecture modeling, hobbies, toys, electronic gadgets, decorative items and jewelry, etc. All these products can be customized according to the consumer's demands simply by altering the digital image and other necessary printing consumables. There are different kinds of technologies associated with 3D printing; they can be based on binder jetting or photopolymer jetting technologies. Although the working principles behind both are different, they demonstrate very efficient and versatile printing work flows. In both technologies, what is very important is the process of the digitalization. Digitalization includes development of 3D objects using computer aided designing (CAD) software. At the beginning, the digital design contains only the surface of the 3D object, using unit shapes which can be multiplied. In the next step, this model is split into multiple steps according to the required printing resolution from bottom to top of the 3D object. These processes are combined by tessellation. Once the process of digitalization is complete, the file is imported to the printer software-recognition platform. In this phase, conversion of the digital image to software-intelligible data is performed. This is critical because this conversion can lead to a loss of data, depending on the printing capability of the machine. Finally, these files are passed to the individual printing technology to develop the 3D printed object. In the case of binder jetting technology, using the inkjet printheads, fluids are jetted onto a bed of powder, and these fuse during UV curing or the thermal post-printing step. The bed is replenished with a flood of new powder and jetting is repeated for the next layer/s. The entire process of powder flooding and jetting is repeated until all the layers are printed from the bottom to the top of the 3D object. In these cases, several different kinds of color are desired, each having different mechanical or aesthetic properties, which can be realized using different inkjet printheads and jetting fluids. When all the layers are printed, the 3D object is extracted from the powder bath and dipped into the infiltration solution. Here the object becomes more mechanically stable, and densification of the intermediate porous layers which form the structure occurs. In the case of photopolymer jetting technology, the same procedure is used without the need for powder, due to the facts that the photopolymer can be UV cured and that structural stability is maintained by the object itself. If a multicolored figure is desired, this can be realized by the use of an increased number of photopolymer fluids, which can be applied inside different numbers of printheads. The realizable printing resolution in the latter case is much better than in the former, due to the non-necessity for filler or materials for supporting the structure. The next section will deal with the market in terms of inkjet technology and its manufacture.

2.3.7.2 Selective View of the Market for Inkjet Technology

The inkjet printing technology industry is enormous and presently there are several competitors in this industry. The world of digital printing technology varies based on the backend technology. Some of the technologies are classic, and others are "state of the art". In inkjet printing technology, companies such as Canon, Fujifilm, Xerox, Agfa, Kodak, Hewlett Packard, etc. are world leaders in the manufacturing of laboratory-to-fabrication inkjet printers. A second class of companies such as Xaar, Konica Minolta, Ricoh, etc. are leaders in printhead manufacturing, developing print solutions or integrations. The biggest technological trade fair for printing and finishing technology took place in May–June 2016. The trade show is called Drupa. Here it could be seen very clearly that the world of printing technology is shifting from conventional roll-to-roll (R2R) analog printing technologies, e.g. offset lithography, gravure, flexography, etc., to high-throughput digital printing technologies for sheet-to-sheet (S2S), roll-to-sheet (R2S), and roll-to-roll (R2R) production processes directed to the application and the requirements of the intended product. The primary parameters differentiating such printing systems are the following: a) printing speed, b) printable area, c) printing resolution, d) color reproduction, e) backend technology and (f) industrial processing capability. Following is a showcase of exemplary inkjet printers and their corresponding technical specifications; these aspects makes them very attractive for graphic art printing press companies.

One of the companies heavily engaged in digital printing technology for graphic arts is Agfa. Agfa has a series of printers called Jeti Titan, based on piezo-electric DoD technology. The printing capability of such printers is enormous, e.g. six-color printing with white on board, printing resolution of 720 dpi × 1200 dpi with drop volume of 7 pL, printable area of 2 m × 3 m, S2S and R2S compatibility and more options for industrial adaptation and usability.

The next company which is very active in this respect is Canon. Canon manufactures various kinds of inkjet printers that suit S2S and R2R platforms, using piezo-electric as well as thermal DoD backend technologies. Some common examples are the Océ Image Stream and VarioPrint Ultra Line series, which have industrial capabilities, e.g. offering up to 2.7 m/s on a 76 cm-wide web, throughput of over 19,200 sheets/h for B2 format (500 mm × 707 mm), 3,500 letter (215.9 mm × 279.4 mm) images per minute, which can be printed with native printing resolution of 1200 dpi × 1200 dpi. The company also produces printing machines based on modified technology such as transfer inkjet technology to achieve high printing speeds and resolutions; this technology uses the transfer of an image developed on an intermediate medium, e.g. blanket, to the required substrate.

One of the other eminent companies active in the market of graphic art printing is Fujifilm. The company produces various kinds of digital presses based on inkjet technology (piezo-electric DoD). One such high-performance R2R or R2S digital press is Fujifilm Jet Press 540 W, which has many capabilities, such as the production of four gray-level values from the same printhead, providing printing resolution of 600 dpi × 600 dpi, offering printing speeds up to 127 m/min, for compatible substrate width of 546 mm and for various substrate thicknesses and grammage, such as 64–157 gm^{-2} paper. The second industrially relevant printing machine is the S2S-compatible Fujifilm X3. The printer is a flatbed production platform which uses the technology of

UV-based inks coupled with a powerful UV flatbed curing station. The print-head installed in this machine offers drop volume between 9–27 pL, which can result into up to 600 dpi × 600 dpi. Production options vary from 40–180 beds/h for a maximum substrate size of 3.22 m × 1.6 m. Another printing machine offered by Fujifilm is called Acuity LED-1600. The printer uses dedicated flatbed design with high-resolution greyscale printheads used for Fujifilm Uvijet UV curing inks extended for 4, 6 or 8 color channels, with standard (1.25 m × 2.5 m), double-sized (3.08 m × 2.5 m) bed versions and large (3.05 m × 2.5 m) bed sizes that can give a throughput of up to 35–155 m^2/h. It has added features such as a multi-zone/dual zone vacuum table and a powerful instant-curing UV system. And finally, the last example is the Fujifilm Uvistar Pro8, which uses the Fujifilm Uvijet UV curing system and parallel drop size (PDS) technology for smooth images, offering 600–1200 dpi printing resolution, and Fujifilm Uvijet QN–UV curing inks for eight colors, which also includes lights and white as options. The throughput of such system is enormous: up to 275–353 m^2/h for a maximum printing width of 3.5–5 m, respectively [75, 77–82].

Even in this era, when digital printing technology for large-volume commercial printed products is needed, collaboration between companies such as Heidelberg AG and Fujifilm, who are masters in offset lithography and inkjet technology respectively, is very predictable. This marks the beginning of a new hybrid technology. An example can be seen with the Heidelberg AG Gallus Labelfire 340 (shown in Figure 2.16b), which is a printer used for producing individualization, e.g. barcodes, data-matrix codes, numbers, etc. The technologies are merged, with resulting capabilities such as printing of UV films with inkjet (piezo-electric DoD) using MEMS based on Samba Technology, production speed up to 50 linear m/min (regardless of number of colors), production throughput of 1,020 m^2/h, printing resolution of 1200 dpi × 1200 dpi with 2 pL drop volume, which gives approximately 2400 dpi × 2400 dpi, up to eight digital color units and print width of maximum 340 mm. Another example is Primefire 106 + L from Heidelberg AG (shown in Figure 2.16a). The printer is again a high-aspect R2R digital printing press using Samba Technology (high width and high printing resolution). Another is RotaJet L from Koenig & Bauer Group AG. The printing system is again based on piezo-electric DoD technology, with native resolution 1200 dpi × 1200 dpi, and uses a stack of printhead bars in an array along with KBA RotaColor polymer pigment ink. The maximum printing speed ranges from 150–300 m/min for a substrate width of 400–1380 mm; the post-processing of the printed films is controlled by infrared dryers combined with hot air blowers on the moving paper web. As a result, print production capability of 87–349 million A4-sized sheets/month is possible with throughput of 3.3–11.8 million m^2/month.

Another interesting market is the field of research and development. Here, much research is focused on tuning the drop volume from the inkjet printhead

(a)

(b)

Figure 2.16 Images of printers (a) Primefire 106 + L from Heidelberg AG and (b) Heidelberg AG Gallus Labelfire 340 [86, 88].

for producing fine lines, graphics and security-based application products. One such technology is Super-fine Inkjet Technology (SIJ) and one of the main producers of SIJ products is SIJTechnology Inc. from the Flexible Electronics Research Center and National Institute of Advanced Industrial Science and Technology, Tsukuba, Ibaraki, Japan. The best-selling product is the SIJ-S050 from SIJTechnology Inc, Figure 2.17. The printer is based on electro-hydrodynamic inkjet technology. The fundamental mechanism behind the generation and tuning of the drop shape and volume was explained in the previous section. The printer offers specialties such as: downscaling of the drop volume from 0.1 fL to 10 pL from the same nozzle; the ability of the printer to work with inks having viscosities of 0.5–10,000 cps (without heating); and the droplets being one-tenth smaller in size and one-thousandth smaller in volume than those that can be used for developing patterns with line resolution < 1 μm. However, there are some drawbacks, such as the inability to upscale to multi-nozzle systems and a printing area of 5 cm × 5 cm. The technology suits a

(a) (b)

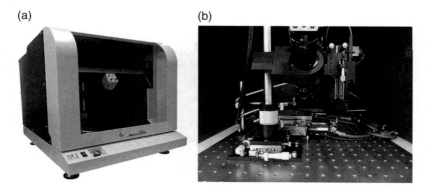

Figure 2.17 SIJ inkjet printer SIJ-S050 from SIJTechnology Inc. (a) full machine, (b) inside view [70, 71]. (*See insert for color representation of the figure.*)

variety of inks and especially non-graphic-based inks such as conductive, insulating, resistant, UV, solvent and protein inks, etc. that are dedicated to printed functionalities and electronics [70, 71].

The next upcoming commercial market of note is the 3D inkjet printing application area, with the development of a fourth dimension which is built using the pre-existing products. An example is the Omnifire printer produced by Heidelberg AG. The printing system can be seen in Figure 2.18. The printer uses the Samba Inkjet Technology (piezo-electric DoD) from Fujifilm and develops printed layers of graphics on round or cylindrical objects with a diameter of 10–300 mm; it allows up to four colors including opaque white, with protection coating as an option, and a single-pass printhead with 360 dpi as native printing resolution. The printer also includes drying or curing systems based on the type of ink used, and a modular mounting system for different objects enabled by four-axis-robotics to move the objects during the printing process [88][90].

2.3.7.3 Advanced Applications: Printed Functionalities and Electronics

Printing beyond colors is something which is gaining attention worldwide. The pull towards printed functionalities and printed electronics is becoming more attractive in laboratory-based research and development, and also in industrialization of such technologies for commercial products. Prominent products and companies in the area are: Autodrop dispensers from microdrop Technologies GmbH, the Dimatix Materials Printer and other industrial inkjet equipment from Fujifilm Dimatix, and inkjet printers from Notion Systems GmbH, Ceradrop and Pixdro. All these inkjet printer manufacturing companies rely on the printhead manufacturers, e.g. Fujifilm Dimatix, Konica Minolta, Ricoh, Xaar, etc. to develop their printing systems. The inkjet printers produced in this case follow the same concept from laboratory to fabrication,

Figure 2.18 Image of application illustrating the potential for printing of graphics on three-dimensional products using inkjet technology, e.g. the Omnifire from Heidelberg AG.

which will promote industrialization of printed functionality and printed products in the market.

As for applications, many products are already being investigated at laboratory scale, with printed functionality products that include printed sieves and printed microfluidics for medical and point-of-care medical diagnostic applications [98, 99]. There are also opportunities for producing inkjet-printed three-dimensional structures on paper for electronic applications [100]. Also, with inkjet technology, development of surfaces for light-absorbing/light-reflecting purposes is possible. One such example is the development of printed lenses, which is not now restricted to laboratory-level research since products such as micro-lenses are industrialized worldwide [101]. Other examples of printed functional structures of great interest include supra-balls and printed photonics which can be produced by inkjet printing technology where the aggregates are self-assembled with special polymer materials [102, 103]. Another exceptionally promising field is printed electronics, where many researchers have introduced materials that can already be used for the fabrication process. This is mainly dedicated to the inkjet printing process and its corresponding applications [71, 72, 104]. Examples include printed conductive tracks which can be used for producing conductive grids for displays or capacitive screens and electronically functional textiles [74–78, 105]. Here the processing of the deposited materials plays a very important role; therefore, fine tuning of the parameters is necessary. Several researchers have carried out investigations in the direction of developing printed active and passive devices, e.g. inkjet-printed thin-film transistors, capacitors, diodes, etc. and have also shown the potential to upscale the fabrication technology for industrialization [79–82, 106–109]. Other examples are investigations in the field of printed radio-frequency identification tags [83, 84, 85 86, 88], energy harnessing through printed solar cells [85, 87], detection or sensing using printed sensors or detectors [88–90], and developing circuits which can be used for various electronic requirements such as signal rectification or filtering and electronic switching [84, 90, 91]. All these applications and upscalable approaches show that inkjet technology has become mature and is capable of being implemented in diverse ways to reach several commercial markets.

2.3.8 Drying Process

The drying process is a fundamental common step of different printing techniques whereby inks in the liquid state are dried in order to form a film and properly adhere to the substrate. Different procedures have been applied in this drying step, including ultraviolet (UV) curing, infrared curing (IR), solvent evaporation at specific temperatures, and a mixture of different procedures [110, 111]. Drying by solvent evaporation is a physical process, the most frequently used as a drying step in printing techniques, and consists of the removal of solvent for the formation of the dried solid [112].

There are different systems for evaporation drying, including air, vacuum, drum and centrifugation drying [113]. The most frequently used systems for evaporation drying are air and vacuum drying or a combination of both. The evaporation of the solvent is affected by temperature, vapor pressure, surface/volume ratio and rate of air flow over the surface [112]. In this case, the drying curves (mass of ink solvent vs time) shows two main stages: a constant rate period and a falling rate period. The constant rate is the most important stage, and is controlled by the fluid dynamic state of the drying air. For electrode inks for printed batteries, the polymer binder crystallizes and redistributes during this process and the particles within the ink rearrange for the formation of the 3D interconnected structure [111, 114].

The falling rate stage is limited by the rate at which the liquid can pass through the solid to the free surface [111].

The mass transfer rate ($\dot{m}_D$) of the drying system is given in the following equation [111]:

$$\dot{m}_D = \frac{\overline{h}}{\rho c_p Le^{2/3}} A_s M (C_s - C_A) \tag{2}$$

Where $\overline{h}$ is the average heat transfer coefficient, ρ is the gas density, c_p is the specific heat at constant pressure, Le is the Lewis number, A_s is the surface area, M is the molar mass, C_s is the solvent vapor concentration of the ink surface and C_A is the solvent vapor concentration of air.

For low concentration of the solvent, the f ρ and c_p are approximately those of pure air.

The difference between air and vacuum systems is that in vacuum systems the environmental pressure is reduced, which lowers the heat needed for rapid drying. For vacuum systems, the solvent evaporation temperature does not need to be as high in comparison to air systems, and the drying is often faster. Typically, vacuum drying shows advantages as less energy is needed for drying, cutting down on the economic and environmental costs and in the potential damage in the materials caused by the drying process [115].

Evaporation rates of the solvent generally show an inverse relationship to the respective boiling point, the vapor pressure value of the solvent having particular relevance to drying systems. Typically, the value of the vapor pressure of solvents increases as the temperature increases. In the case where the solvent pressure value is 760 mmHg (atmospheric pressure), solvent evaporation occurs throughout the liquid mass and not only at the surface [116].

For ultraviolet (UV) and infrared (IR) curing, the inks are cured under exposure to different energy types, i.e., UV light and infrared radiation, respectively [110]. In the case of UV drying, UV radiation with wavelengths between 100 and 380 nm initiates a radiation polymerization or crosslinking (chemical drying), transforming the liquid ink into a solid film.

For this system, control of the physical properties of the ink is very important, since the viscosity of inks limits the choice of the molecular weight of the monomers, solids and cure rate [110].

The advantages of this system are the lack of emission of volatile organic compound, the lower energy requirement, the space saving of the curing equipment, good adhesion to substrates, and overall increased "environmental friendliness" [110].

As well as these systems, it is also possible to cure the inks through the use of electron-beam [117].

2.3.9 Process Chain

The manufacturing process of printed batteries requires different steps and procedures.

The main procedures for the fabrication of printed batteries are shown in Figure 2.19. For obtaining functional layers for the different components of the batteries it is first necessary to use suitable inks appropriate to each printing technique. For preparation of the ink, fabrication involves mixing of the different constituents (polymer binder, active material and conductive agent) that inks for electrodes require, whereas for the fabrication of the layers, coating, pressing, slitting and drying are often needed.

Each electrode and separator/electrolyte can be produced separately, the battery assembly process being then a critical step. In the case of printed batteries, the sealing or encapsulation process is essential for keeping the atmosphere constant without CO_2 inside the battery during its lifetime. This process consists in applying a sealing layer based on a polymer glue that can be processed by the application of heat or pressure [7].

It is also necessary to have good leak tightness for the electrolyte solution.

The substrate should present excellent barrier properties and must be stable to thermal treatment at the temperatures (approximately 130 °C) needed to increase the electric conductivity of the current collectors and electrodes.

One advantage of printed batteries is their integration into the devices for different applications, which is a differentiating factor with respect to other battery types, as well as other properties such as capacity, weight and dimensions compared to conventional batteries.

Another important procedure is the testing step applied to all batteries, which results in the activation process and the solid electrolyte interface (SEI)-layer formation on the anode and cathode surface. This procedure facilitates charge/discharge behavior, affecting battery performance, life and safety.

From a practical point of view

In a production line the abovementioned processes need to be combined to enable the production of batteries. The approach of combining different technologies for the application of layers with varying parameters, such as thickness or density, is common in a wide spectrum of industries but is exceptional in the printing industry.

Setting up a production line accordingly means the careful optimization of various characteristics such as footprint, productivity, yield, operating time and cost of operation. The integration of all necessary process steps is a straight

Figure 2.19 Main procedures for printed batteries.

approach in minimizing the required footprint, material handling times, and, eventually, equipment costs.

However, it must be taken into account that each of the process steps has different productivity, meaning optimal web speed. This is mainly associated with the required length of the dryer, which can be adapted for each process step to a target web speed. Nonetheless, most printing processes have a speed window optimal for the application of the ink/paste, due to its physical properties. It can thus be sensible to chain only a few processes and run the fastest and/or slowest processes in dedicated standalone tools when optimizing the production costs in relation to the targeted annual production of batteries.

Because printed batteries are not yet fully commercialized, it is also recommended to follow the idea of lab-to-fab by scaling production with the development of the battery. When setting up a pilot line it can be beneficial to start with one production tool incorporating all process technologies, and later on, e.g. upon arrival of a certain milestone, extend production capacities by either purchasing new equipment or upgrading existing equipment.

Furthermore, it must be taken into account that the functional materials for batteries will improve over time and likely substitute existing materials. However, it may be that the improved materials show other properties, thus requiring a different printing technology for the application. In these cases it is of the utmost importance that the process chain can be extended by the required process step or that substitution of the now superfluous process technologies can take place.

2.3.10 Printing of Layers

The layers of the different components (current collectors, anode and cathode electrodes and separators) in printed batteries can be produced by different printing techniques using suitable inks with required viscosity. Table 2.4 shows different printing techniques reported in the literature for manufacturing specific components of printed batteries. The references shown in Table 2.4 for each printing technique are representative.

The current collector layers are based on conductive inks based on carbonaceous fillers such as carbon nanotubes (CNT) and carbon nanofibers (CNF), metallic inks of silver, zinc or copper metals deposited on flexible substrates and conductive fabrics [26].

The most frequently printed components of printed batteries are the electrodes, which have been printed using different active materials. Independently of the active material, the literature shows that it is possible to produce the electrodes by different printing techniques [7].

The battery component that requires extra attention and development is the separator, as it must show high ionic conductivity and high thermal and mechanical stability [7]. For printed batteries, the separator layer is

Table 2.4 Printing techniques applied to printed batteries.

Printing technique	Battery type	Component	Ref.
Screen printing	Lithium-ion	Electrodes: $Li_4Ti_5O_{12}$	[9]
Inkjet printing	Lithium-ion	Electrodes: SnO_2	[118]
Doctor blade	Lithium-ion	Electrodes: $Li_4Ti_5O_{12}$	[119]
Spray-painting technique	Lithium-ion	Electrodes: $Li_4Ti_5O_{12}$ and $LiCoO_2$	[120]
3D-printing	Lithium-ion	Electrodes: $Li_4Ti_5O_{12}$ and $LiFePO_4$	[121]
Stencil-printing technique	Zn/MnO$_2$	Electrodes: Zn and MnO_2	[6]
Flexography technique	Zn/MnO$_2$	Electrodes: Zn and MnO_2	[122]
Screen printing	Zn/MnO$_2$ (primary)	Electrodes: Zn and MnO_2	[123]
Mechanically imprinted with a maze-patterned PDMS stamp	Lithium-ion	Separator	[124]

typically a polymer gel electrolyte with different constituents, including salts or ionic liquids [7].

More details about the printing layers for the different components are presented in the following chapters of this book.

2.4 Conclusions

Numerous printing techniques have been established. Some, such as screen printing or coating technologies, are ideal for the deposition of homogenous functional layers of defined layer thickness in large areas. These are technologies that work on web-based substrates enabling large area throughput.

When dealing with small and sophisticated batteries, digital printing technologies such as dispensing or inkjet printing are appropriate technologies that may limit the particle sizes contained in the processed inks.

Acknowledgements

Fraunhofer ENAS and Chemnitz University of Technologies have received funding for research and development of printed batteries using printing technologies in the following projects: BatMat (BMBF 13N11400), leiTEX (BMBF Zwanzig20 03ZZ0616B), and SIMS (FP7 GA No. 257372). The following staff members were involved in the development: André Kreutzer, Ulrike Geyer, Michael Espig and Monique Helmert.

This work was supported by the Portuguese Foundation for Science and Technology (FCT) in the framework of Strategic Funding UID/FIS/04650/2013, project PTDC/CTM-ENE/5387/2014 and grant SFRH/BPD/112547/2015 (C.M.C.). The authors thank the Basque Government Industry Department for its financial support under the ELKARTEK Program.

References

1 Park, J.K. (2012) *Principles and Applications of Lithium Secondary Batteries*, Wiley-VCH, Weinheim.

2 Hoerber, J., Glasschroeder, J., Pfeffer, M., Schilp, J., Zaeh, M., Franke, J. (2014) Approaches for additive manufacturing of 3D electronic applications. *Procedia CIRP* **17**, 806–811.

3 Pollet, B.G., Staffell, I., Shang, J.L. (2012) Current status of hybrid, battery and fuel cell electric vehicles: from electrochemistry to market prospects. *Electrochimica Acta* **84**, 235–249.

4 Rydh, C.J., Svärd, B. (2003) Impact on global metal flows arising from the use of portable rechargeable batteries. *Science of the Total Environment* **302**, 167–184.

5 Thackeray, M.M., Wolverton, C., Isaacs, E.D. (2012) Electrical energy storage for transportation-approaching the limits of, and going beyond, lithium-ion batteries. *Energy & Environmental Science* **5**, 7854–7863.

6 Gaikwad, A.M., Steingart, D.A., Ng, T.N., Schwartz, D.E., Whiting, G.L. (2013) A flexible high potential printed battery for powering printed electronics. *Applied Physics Letters* **102(23)**, 233302.

7 Sousa, R.E., Costa, C.M., Lanceros-Méndez, S. (2015) Advances and future challenges in printed batteries. *ChemSusChem* **8**, 3539–3555.

8 Iwakura, C., Fukumoto, Y., Inoue, H., Ohashi, S., Kobayashi, S., Tada, H., Abe, M. (1997) Electrochemical characterization of various metal foils as a current collector of positive electrode for rechargeable lithium batteries. *J. Power Sources* **68**, 301–303.

9 Prosini, P.P., Mancini, R., Petrucci, L., Contini, V., Villano, P. (2001) Li4Ti5O12 as anode in all-solid-state, plastic, lithium-ion batteries for low-power applications. *Solid State Ionics* **144(1–2)**, 185–192.

10 Gören, A., Mendes, J., Rodrigues, H.M., Sousa, R.E., Oliveira, J., Hilliou, L. *et al.* (2016) High performance screen-printed electrodes prepared by a green solvent approach for lithium-ion batteries. *J. Power Sources* **334**, 65–77.

11 Choi, M.G., Kim, K.M., Lee, Y.-G. (2010) Design of 1.5 V thin and flexible primary batteries for battery-assisted passive (BAP) radio frequency identification (RFID) tag. *Current Applied Physics* **10**(4, Supplement) e92–e96.

12 Gaikwad, A.M., Whiting, G.L., Steingart, D.A., Arias, A.C. (2011) Highly flexible, printed alkaline batteries based on mesh-embedded electrodes. *Adv. Mater.* **23**, 3251–3255.

13 Sousa, R.E., Oliveira, J., Gören, A., Miranda, D., Silva, M.M., Hilliou, L. *et al.* (2016) High performance screen printable lithium-ion battery cathode ink based on C-LiFePO4. *Electrochimica Acta* **196**, 92–100.

14 Lee, J.-H., Wee, S.-B., Kwon, M.-S., Kim, H.-H., Choi, J.-M., Song, M.S. *et al.* (2011) Strategic dispersion of carbon black and its application to ink-jet-printed lithium cobalt oxide electrodes for lithium ion batteries. *J. Power Sources* **196**, 6449–6455.

15 Braam, K.T., Volkman, S.K., Subramanian, V. (2012) Characterization and optimization of a printed, primary silver–zinc battery. *J. Power Sources* **199**, 367–372.

16 Ho, C.C., Murata, K., Steingart, D.A., Evans, J.W., Wright, P.K. (2009) A super ink jet printed zinc–silver 3D microbattery. *J. Micromech. Microeng.* **19(9)**, 094013.

17 Delannoy, P.E., Riou, B., Lestriez, B., Guyomard, D., Brousse, T., Le Bideau, J. (2015) Toward fast and cost-effective ink-jet printing of solid electrolyte for lithium microbatteries. *J. Power Sources* **274**, 1085–1090.

18 Willert, A., Killard, A.J., Baumann, R.R. (2014) Tailored printed primary battery system for powering a diagnostic sensor device. *J. Print Media Technol. Res.* **3**, 57–64.

19 Wong, W.S., Salleo, A. (2009) *Flexible Electronics: Materials and Applications*, Springer, New York, NY.

20 Chang, J., Ge, T., Sanchez-Sinencio, E. (2012) Challenges of printed electronics on flexible substrates. *2012 IEEE 55th International Midwest Symposium on Circuits and Systems (MWSCAS)* 582–585.

21 Dearden, A.L., Smith, P.J., Shin, D.-Y., Reis, N., Derby, B., O'Brien, P. *et al.* (2005) Curing temperature silver ink for use in ink-jet printing and subsequent production of conductive tracks. *Macromolecular Rapid Communications* **26**, 315–318.

22 van Osch, T.H.J., Perelaer, J., de Laat, A.W.M., Schubert, U.S. (2008) Inkjet printing of narrow conductive tracks on untreated polymeric substrates. *Adv. Mater.* **20**, 343–345.

23 Bonadiman, R., Marques, M., Freitas, G., Reinikainen, T. (2008) Evaluation of printing parameters and substrate treatment over the quality of printed silver traces. *2008 2nd Electronics System-Integration Technology Conference*, 1343–1348.

24 Madeij, E., Espig, M., Baumann, R.R., Schuhmann, W., Mantia, F.L. (2014) Optimization of primary printed batteries based on Zn/MnO2. *J. Power Sources* **261**, 356–362.

25 Kipphan, H. (2001) *Handbook of Print Media: Technologies and Production Methods*, Springer, Berlin.

26 Gaikwad, A.M., Arias, A.C., Steingart, D.A. (2015) Recent progress on printed flexible batteries: mechanical challenges, printing technologies, and future prospects. *Energy Technology* **3**, 4305–4328.

27 Galagan, Y., Rubingh, J.-E.J., Andriessen, R., Fan, C.-C., Blom, P.W., Veenstra, S.C. *et al.* (2011) ITO-free flexible organic solar cells with printed current collecting grids. *Sol. Energy Mater. Sol. Cells* **95**, 1339–1343.

28 Krebs, F.C., Jørgensen, M., Norrman, K., Hagemann, O., Alstrup, J., Nielsen, T.D., *et al.* (2009) A complete process for production of flexible large area polymer solar cells entirely using screen printing—first public demonstration. *Sol. Energy Mater. Sol. Cells* **93**, 422–441.

29 Shin, D.-Y., Lee, Y., Kim, C.H. (2009) Performance characterization of screen printed radio frequency identification antennas with silver nanopaste. *Thin Solid Films* **517**, 6112–6118.

30 Siden, J., Nilsson, H.-E. (2007) Line width limitations of flexographic-screen- and inkjet printed RFID antennas. In *Antennas and Propagation Society International Symposium, 2007*, IEEE, 1745–1748.

31 Jabbour, G., Radspinner, R., Peyghambarian, N. (2001) Screen printingfor the fabrication of organic light-emitting devices. *IEEE Journal on Selected Topics in Quantum Electronics* **7**, 799–773.

32 Fouchal, F., Dickens, P. (2007) Adaptive screen printing for rapid manufacturing. *Rapid Prototyping Journal* **13**, 284–290.

33 Riemer, D.E. (1988) *Ein Beitrag zur Untersuchung der physikalisch-technischen Grundlagen des Siebdruckverfahrens*, Technische Universität Berlin, Berlin, 95.

34 Wendler, M., Hübner, G., Krebs, M. (2011) Development of printed thin and flexible batteries. *International Circular of Graphic Education and Research* **4**, 32–41.

35 Xu, Y., Schwab, M.G., Studwick, A.J., e.a. Hennig, I. (2013) Screen-Printable Thin film supercapacitor device utilizing graphene/polyaniline inks. *Adv. Energy Mater.* **3**, 1035–1040.

36 Kang, K.-Y., Lee, Y.-G., Shin, D.O., Kim, J.-C., Kim, K.M. (2014) Performance improvents of pouchtype flexible thin-film lithium-ion batteries by modifying sequential screen-printing process. *Electrochimica Acta* **138**, 294–301.

37 Hilder, M., Winter-Jensen, B., Clark, N.B. (2009) Paper-based, printed zinc-air battery. *J. Power Sources* **194**, 1135–1141.

38 Kazani, I., Hertleer, C., Mey, G.D. (2012) Electrical conductive textiles obtained by screen printing. *Fibers and Textiles in Eastern Europe* **20**, 57–63.

39 Riemer, D.E. (1989) The theoretical fundamentals of the screen printing process, *Microelectronics International* **6(1)**, 8–17.

40 Foster, C.W. (2016) *Fundamentals of Screen-Printing Electrochemical Architektures*, Springer-Briefs in Applied Science and Technology, Cham, Switzerland.

41 Brown, D.O. (1986) Screen printing – an integrated system. *International School of Hydrocarbon Measurement (ISHM) Proceedings* **586**.

42 Phail, D.M. (1996) Screen printing is a science, not an art. *Soldering and Surface Mount Technology* **8(2)**, 25–28.

43 Pan, J., Tonkay, G., Quintero, A. (1999) Screen printing process design of experiments for fine line printing of thick film ceramic substrates. *J. Electronics Manufacturing* **9**, 203–213.

44 Willfahrt, A., Stephens, J., Hübner, G. (2011) Optimised stencil thickness and ink film deposition. *International Circular of Educational Institutes for Graphic Arts* **4**, 6–17.

45 Kuo, H.-P., Yang, C.-F., Huang, A.-N., Wu, C.-T., Pan, W.-C. (2014) Preparation of the working electrode of dye-sensitized solar cells: effects of screen printing parameters. *J. Taiwan Inst. Chem. Eng.* **45**, 2340–2345.

46 Schmidt, G.C., Bellmann, M., Meier, B., Hambsch, M., Reuter, K., Kempa, H. *et al.* (2010) Modified mass printing technique for the realization of source/drain electrodes with high resolution. *Organic Electronics* **11**, 1683–1687.

47 Huebner, C.F., Carroll, J.B., Evanoff, D.D., Ying, Y., Stevenson, B.J., Lawrence, J.R. *et al.* (2008) Electroluminescent colloidal inks for flexographic roll-to-roll printing. *J. Mater. Chem.* **18**, 4942–4948.

48 Krebs, F.C., Fyenbo, J., Jørgensen, M. (2010) Product integration of compact roll-to-roll processed polymer solar cell modules: methods and manufacture using flexographic printing, slot-die coating and rotary screen printing. *J. Mater. Chem.* **20**, 8994–9001.

49 Mäkelä, T., Jussila, S., Kosonen, H., Bäcklund, T., Sandberg, H., Stubb, H. (2005) Utilizing roll-to-roll techniques for manufacturing source-drain electrodes for all-polymer transistors. *Synthetic Metals* **153(1–3)**, 285–288.

50 Jung, M., Kim, J., Noh, J., Lim, N., Lim, C., Lee, G. *et al.* (2010) All-printed and roll-to-roll-printable 13.56-MHz-operated 1-bit RF tag on plastic foils. *IEEE Transactions on Electron Devices* **57**, 571–580.

51 Sung, D., de la Fuente Vornbrock, A., Subramanian, V. (2010) Scaling and optimization of gravure-printed silver nanoparticle lines for printed electronics. *IEEE Transactions on Components and Packaging Technologies* **33(1)**, 105–114.

52 Pauliac-Vaujour, E., Stannard, A., Martin, C., Blunt, M.O., Notingher, I., Moriarty, P. *et al.* (2008) Fingering instabilities in dewetting nanofluids. *Physical Review Letters* **100**, 176102.

53 Sullivan, T.M., Middleman, S. (1986) Film thickness in blade coating of viscous and viscoelastic liquids. *J. Non-Newtonian Fluid Mech.* **21(1)**, 13–38.

54 Pham, V.H., Cuong, T.V., Hur, S.H., Shin, E.W., Kim, J.S., Chung, J.S. *et al.* (2010) Fast and simple fabrication of a large transparent chemically-converted graphene film by spray-coating. *Carbon* **48**, 1945–1951.

55 Yang, H., Jiang, P. (2010) Large-scale colloidal self-assembly by doctor blade coating. *Langmuir* **26**, 13173–13182.

56 Berni, A., Mennig, M., Schmidt, H. (2004) Doctor blade. In M.A. Aegerter, M. Mennig (eds) *Sol-Gel Technologies for Glass Producers and Users*, Springer, Boston, MA, 89–92.

57 Pudas, M., Hagberg, J., Leppävuori, S. (2004) Printing parameters and ink components affecting ultra-fine-line gravure-offset printing for electronics applications. *J. European Ceramic Soc.* **24**, 2943–2950.

58 Hoth, C.N., Schilinsky, P., Choulis, S.A., Brabec, C.J. (2008) Printing highly efficient organic solar cells. *Nano Letters* **8**, 2806–2813.

59 Krebs, F.C. (2009) Fabrication and processing of polymer solar cells: A review of printing and coating techniques. *Sol. Energy Mater. Sol. Cells* **93**, 394–412.

60 Tok, A.I.Y., Boey, F.Y.C., Khor, M.K.A. (1999) Tape casting of high dielectric ceramic substrates for microelectronics packaging. *J. Mater. Eng. Perform.* **8**, 469–472.

61 Yu, B.K., Vak, D., Jo, J., Na, S.I., Kim, S.S., Kim, M.K. *et al.* (2010) Factors to be considered in bulk heterojunction polymer solar cells fabricated by the spray process. *IEEE J. Sel. Top. Quantum Electron.* **16**, 1838–1846.

62 Ligneel, E., Lestriez, B., Hudhomme, A., Guyomard, D. (2007) Effects of the solvent concentration (solid loading) on the processing and properties of the composite electrode. *J. Electrochem. Soc.* **154**(3), A235–A241.

63 Hutchings, I.M. (2013) *Fundamentals of Inkjet Technology*, John Wiley and Sons, Ltd, Chichester.

64 Wijshoff, H.M.A. (2010) The dynamics of the piezo inkjet printhead operation. *Physics Reports* **491**(4–5), 77–177.

65 Reis, N. (2005) Ink-jet delivery of particle suspensions by piezoelectric droplet ejectors. *Applied Physics* **97**, 094903/1–094903/6.

66 Jang, D. (2009) Influence of fluid physical properties on ink-jet printability. *Langmuir* **25**, 2629–2635.

67 Murata, K. (2005) Super-fine ink-jet printing: toward the minimal manufacturing system. *Microsystem Technology* **12**, 2–7.

68 Xiong, Z. (2011) The application of inkjet direct writing in solar cell fabrication: an overview. *International Conference on Electronic Packaging Technology & High Density Packaging*.

69 Murata, K. (2003) Super-fine ink-jet printing for nanotechnology. *Proceedings of the International Conference on MEMS, NANO and Smart Systems (ICMENS'03)*.

70 Laurila, M.-M. (2015) Inkjet printed single layer high-density circuitry for a MEMS device. *IEEE – Electronic Components & Technology Conference* 978–1–4799–8609–5 968–972.

71 Kirchmeyer, S. (2005) Scientific importance, properties and growing applications of poly(3,4-ethylenedioxythiophene). *J. Mater. Chem.* **15**, 2077–2088.

72 Singh, M. (2010) Inkjet Printing—Process and Its Applications. *J. Adv. Mater.* **22**, 673–685.

73 Agfa Jeti Titan S / HS. <https://www.agfagraphics.com/global/en/product-finder/jeti-titan-s-hs.html>.

74 Weise, D. (2014) *Conductivity and Microstructure of Inkjet-Printed Silver Tracks Depending on the Digital Pattern: Sintering Process, Substrate and Ink*, Materials Research Society, Boston, MA.

75 Weise, D. (2015) *Intense Pulsed Light Sintering of Inkjet Printed Silver Nanoparticle Ink: Influence of Flashing Parameters and Substrate*, Materials Research Society, Boston, MA.

76 Sowade, E. (2015) Roll-to-roll infrared (IR) drying and sintering of an inkjet-printed silver nanoparticle ink within 1 second. *J. Mater. Chem. C* **3**, 11815–11826.

77 Niittynen, J. (2014) Alternative sintering methods compared to conventional thermal sintering for inkjet printed silver nanoparticle ink. *J. Thin Solid Films* **556**, 452–459.

78 Mitra, K. (2016) Inkjet-printing of conductive tracks on non-woven flexible textile fabrics for wearable applications. *Flexible and Wearable Electronics: Design and Fabrication Techniques*, United Scholars Publications, Cincinnati, OH.

79 Sowade, E. (2016) Up-scaling of the manufacturing of all-inkjet-printed organic thin-film transistors: Device performance and manufacturing yield of transistor arrays. *J. Organic Electronics* **30**, 237–246.

80 Mitra, K. (2017) Inkjet printed metal insulator semiconductor (MIS) diodes for organic and flexible electronic application. *J. Flexible Printed Electron.* **2**, (015003) 1–10.

81 Mitra, K. (2015) Potential up-scaling of inkjet-printed devices for logical circuits in flexible electronics. *International Conference and Exhibition on Nanotechnologies and Organic Electronics (Nanotexnology 2014), Thessaloniki*, 106–114.

82 Mitra, K. (2015) *Infra-red Curing Methodology for Roll-to-Roll (R2R) manufacturing of Conductive Electrodes through Inkjet Technology Applicable for Devices in the Field of Flexible Electronics*, Materials Research Society, San Francisco, CA, 1–6.

83 Cook, B. (2013) Multi-Layer RF capacitors on flexible substrates utilizing inkjet printed dielectric polymers. *IEEE Microwave and Wireless Components Letters* **23**, 353–355.

84 Sternkiker, C. (2016) Upscaling of the inkjet printing process for the manufacturing of passive electronic devices. *IEEE Transactions on Electron Devices* **63**, 426–431.

85 Eggenhuisen, T.M. (2015) High efficiency, fully inkjet printed organic solar with freedom of design. *J. Mater. Chem. A* **3**, 7255–7262.

86 McKerricher, G. (2015) Fully inkjet printed RF inductors and capacitors using polymer dielectric and silver conductive ink with through vias. *IEEE Transactions on Electron Devices* **62**, 1002–1009.

87 Xerxes Steirer, K. (2009) Ultrasonically sprayed and inkjet printed thin film electrodes for organic solar cells. *J. Thin Solid Films* **517**, 2781–2786.

88 Molina-Lopez, F., Briand, D., de Rooij, N.F. (2012) All additive inkjet printed humidity sensors on plastic substrate. *J. Sensors and Actuators B* **166–167**, 212–222.

89 Lorwongtragool, P. (2014) A novel wearable electronic nose for healthcare based on flexible printed chemical sensor array. *J. Sensors* **14**, 19700–19712.

90 Chen, B. (2003) All-polymer RC filter circuits fabricated with inkjet printing technology. *J. Solid-State Electronics* **47**, 841–847.

91 Fukuda, K. (2014) Fully-printed high-performance organic thin-film transistors and circuitry on one-micron-thick polymer films. *Nature Communications* **5**, 1–8.

92 Homepage Notion Systems. <http://www.notion-systems.com/>.

93 Homepage Ceradrop. <http://www.ceradrop.com/en/>.

94 Homepage Pixdro. <http://www.pixdro.com/>.

95 Homepage Konica Minolta. <https://www.konicaminolta.com/inkjet/inkjethead/>.

96 Homepage Ricoh. <http://www.rpsa.ricoh.com/>.

97 Homepage Xaar. <http://www.xaar.com/en/products>.

98 Abe, K. (2010) Inkjet-printed paperfluidic immuno-chemical sensing device. *J. Analytical and Bioanalytical Chemistry* **398**, 885–893.

99 Hammerschmidt, J. (2012) Inkjet printing of reinforcing patterns for the mechanical stabilization of fragile, polymeric microsieves. *Langmuir* **28**, 3316–3321.

100 Nge, T. (2013) Electrical functionality of inkjet-printed silver nanoparticle conductive tracks on nanostructured paper compared with those on plastic substrates. *J. Mater. Chem. C* **1**, 5235–5243.

101 Fakhfouri, V. (2008) Inkjet printing of SU-8 for polymer-based MEMS: a case study for microlenses. *21st International Conference on Micro Electro Mechanical Systems IEEE, Wuhan, China*, 407–410.

102 Sowade, E. (2012) In-flight inkjet self-assembly of spherical nanoparticle aggregates. *Adv. Eng. Mater.* **14**, 98–100.

103 Belgardt, C. (2013) Inkjet printing as a tool for the patterned deposition of octadecylsiloxane monolayers on silicon oxide surfaces. *Phys. Chem. Chem. Phys.* **15**, 7494–7504.

104 Calvert, P. (2001) Inkjet printing for materials and devices. *Chem. Mater.* **13**, 3299–3305.

105 Perelaer, J. (2012) Roll-to-roll compatible sintering of inkjet printed features by photonic and microwave exposure: from non- conductive ink to 40% bulk silver conductivity in less than 15 seconds. *J. Adv. Mater.* **24**, 2620–2625.

106 Mitra, K. (2016) Time-efficient curing of printed dielectrics via infra-red suitable to S2S and R2R manufacturing platforms for electronic devices. *IEEE Transaction on Electron Devices* **63**, 2777–2784.

107 Fukuda, K. (2013) Profile control of inkjet printed silver electrodes and their application to organic transistors. *ACS Appl. Mater. Inter.* **5**, 3916–3920.

108 Medeiros, M. (2013) Inkjet-printed organic electronics: operational stability and reliability issues. *J. ECS Transactions* **53(26)**, 1–10.

109 Mitra, K. (2016) Influence of the process workflows on the electrical properties of a UV-curable polymeric dielectric for all-inkjet-printed capacitors. *J. Wearable Flexible Electron.* **1**(1), 1–10.

110 Hancock, A., Lin, L. (2004) Challenges of UV curable ink-jet printing inks – a formulator's perspective. *Pigment & Resin Tech.* **33**, 280–286.

111 Avcı, A., Can, M., Etemoğlu, A.B. (2001) A theoretical approach to the drying process of thin film layers. *Appl. Therm. Eng.* **21**, 465–479.

112 Wypych, G. (2001) *Handbook of Solvents*, ChemTec, Ontario.

113 Kern, W., Vossen, J.L. (2012) *Thin Film Processes II*, Elsevier Science.

114 Gören, A., Cíntora-Juárez, D., Martins, P., Ferdov, S., Silva, M.M., Tirado, J.L. *et al.* (2016) Influence of solvent evaporation rate in the preparation of carbon-coated lithium iron phosphate cathode films on battery performance. *Energy Technology* **4**, 573–582.

115 Parikh, D.M. (2015) Vacuum drying: basics and application. *Chem. Eng.* **122**(4), 48.

116 Wicks, Z.W., Jones, F.N., Pappas, S.P., Wicks, D.A. (2007) *Organic Coatings: Science and Technology*, John Wiley & Sons, Hoboken, NJ.

117 Likavec, W.R., Bradley, C.R. (1999) Ultraviolet and electron beam radiation curable fluorescent printing ink concentrates and printing inks. Google Patents.

118 Zhao, Y., Zhou, Q., Liu, L., Xu, J., Yan, M., Jiang, Z. (2006) A novel and facile route of ink-jet printing to thin film SnO2 anode for rechargeable lithium ion batteries. *Electrochimica Acta* **51**, 2639–2645.

119 Hu, L., Wu, H., La Mantia, F., Yang, Y., Cui, Y. (2010) Thin, flexible secondary li-ion paper batteries, *ACS Nano* **4**, 5843–5848.

120 Singh, N., Galande, C., Miranda, A., Mathkar, A., Gao, W., Reddy, A.L.M. (2012) Paintable battery. *Scientific Reports* **2**, 481.

121 Sun, K., Wei, T.-S., Ahn, B.Y., Seo, J.Y., Dillon, S.J., Lewis, J.A. (2013) 3D Printing of interdigitated li-ion microbattery architectures. *Adv. Mater.* **25**, 4539–4543.

122 Wang, Z., Winslow, R., Madan, D., Wright, P.K., Evans, J.W., Keif, M. *et al.* (2014) Development of MnO2 cathode inks for flexographically printed rechargeable zinc-based battery. *J. Power Sources* **268**, 246–254.

123 Willert, A., Hammerschmidt, J., Baumann, R.R. (2012) *Mass Printing Technologies for Technical Applications*, Scientific Papers of the University of Pardubice 17 (Series A).

124 Kil, E.-H., Choi, K.-H., Ha, H.-J., Xu, S., Rogers, J.A., Kim, M.R. *et al.* (2013) Imprintable, bendable, and shape-conformable polymer electrolytes for versatile-shaped lithium-ion batteries. *Adv. Mater.* **25**, 1395–1400.

3

The Influence of Slurry Rheology on Lithium-ion Electrode Processing

Ta-Jo Liu[1], Carlos Tiu[2], Li-Chun Chen[1,3] and Darjen Liu[1,3]

[1] *Department of Chemical Engineering, National Tsing Hua University, Hsinchu, Taiwan*
[2] *Department of Chemical Engineering, Monash University, Clayton, Australia*
[3] *Material and Chemical Research Laboratories, Industrial Technology Research Institute, Hsinchu, Taiwan*

3.1 Introduction

Lithium-ion batteries (LIBs) have been used in many consumer products and electric vehicles in recent years. Mixing equipment and mixing processes of the electrode production will directly affect the manufacturing process and performance of batteries. Lithium-ion batteries produced from electrodes with well-dispersed particle distribution have been shown to have significantly improved electrochemical performance and battery life [1–12].The electrode slurry is a solid–liquid mixture comprising active materials, conductive particles, polymer binder and solvent medium. It has a high solid content which consists of nano- and micron-sized particles, and is difficult to disperse in a highly viscous medium. Hence, good mixing of various components in an electrode slurry is an essential processing step in achieving a high quality product. Since the slurry is black and opaque, particle distribution is difficult to observe with confidence using optical devices. An indirect method of assessing the stability and uniformity of the electrode slurry is by rheological means. A good electrode slurry must have the following characteristics: low sedimentation and agglomeration during storage; uniform particle distribution in the slurry and good fluidity to facilitate slurry coating; and high viscosity to prevent sedimentation of particles and binder migration during drying.

The main focus of this chapter is the rheological behavior of lithium-ion electrode slurries, as it is an important processing parameter in every stage of the electrode manufacturing process. It starts by discussing the formulation

Printed Batteries: Materials, Technologies and Applications, First Edition.
Edited by Senentxu Lanceros-Méndez and Carlos Miguel Costa.

and preparation of the slurries, followed by coating of slurries to produce wet electrode films and finally drying of electrodes. The rheological behavior of slurry and its influence on various processing stages are discussed here.

3.2 Slurry Formulation

The first step of the lithium-ion battery electrode manufacturing process is mixing the slurry. Electrode slurry is a multi-component, complex solid–liquid mixture which consists of active materials, conductive particles, polymer binder and solvent medium. The slurry is coated onto a metal current collector substrate and dried to make the electrode layer.

Table 3.1 summarizes various components commonly used for lithium-ion battery electrodes [1–32]. In slurry formulation, the major ingredients are the active materials and the conductive agents. The active material, consisting of complex lithium compound, allows the electrochemical reaction to proceed. The conductive agent improves the electronic conductivity and increases electricity transfer in the electrode. The most commonly used active materials in lithium batteries for the cathode are $LiCoO_2$, $LiNiO_2$, $LiMn_2O_4$, $LiFePO_4$, and $LiNi_xM_yCo_{1-x-y}O_2$; and for the anode are graphite and Si-C. The solid contents of both active materials are over 70wt% [1–32].

Electrode slurries can be classified into aqueous and organic systems according to the type of solvent used. The solvent is an important component of the formulation for controlling the slurry dispersion. Deionized water is normally used for the aqueous system and NMP (N-methyl-2-pyrrolidone) for the organic-based system. The solid content and the viscosity of the slurry are dependent on the amount of solvent added.

Table 3.1 The main ingredients in electrode slurries.

Type	Material	Ratio
Active material	Cathode: $LiCoO_2$, $LiNiO_2$, $LiMn_2O_4$, $LiFePO_4$, $LiNi_xM_yCo_{1-x-y}O_2$ Anode: Graphite, Si/Si-C	80–97wt% (Pure solid weight ratio)
Binder	PVDF, SBR, CMC, PMMA	1–10wt% (Pure solid weight ratio)
Conductive agent	Graphite, Carbon black, Acetylene black	0.5–10wt% (Pure solid weight ratio)
Solvent	NMP, Deionized water	Added content depend on mixing and coating process

A polymeric thickener such as CMC (carboxymethyl cellulose) may be required for the aqueous-based slurry system in order to increase the slurry viscosity and to minimize particle sedimentation. The most commonly used binder for electrode slurry is SBR (styrene-butadiene rubber) for aqueous and PVDF (polyvinylidene fluoride) for organic systems. The binder provides the adhesion strength between particle/particle and particle/current collector. The content of the binder in the electrode slurry is between 1 and 10wt%. A summary of the main ingredients of electrode slurry is given in Table 3.1.

3.3 Rheological Characteristics of Electrode Slurry

Rheological characterization of electrode slurry is essential at various stages of battery processing. For suspensions, the most important rheological property in lithium electrode processing is the viscosity. It directly influences the behavior of the slurry formulation during mixing, production of wet electrode through coating and subsequent drying. Other rheological properties such as viscoelasticity and yield stress are related to the molecular and internal structure of the electrode. Viscosity is used as a means of determining the uniformity of particle dispersion during slurry formulation, the stability of the wet film during coating, and the extent of solvent evaporation during drying. In general, electrode slurry is a non-Newtonian complex mixture which contains a large number of particles of various sizes and shapes, polymeric binder and water or organic solvent with thickener. The three main rheological characteristics of electrode slurries in terms of shear-viscosity, viscoelasticity, and yield stress and their roles at various stages of processing are briefly discussed here.

3.3.1 Viscosity and Shear-Thinning

Viscosity is an important physical property of a fluid, which determines the resistance to flow and fluid deformation, and is defined as the ratio of shear stress over shear rate. In a solid–liquid suspension, it is closely related to the solid concentration of the mixture. For a non-interactive dilute suspension, the suspension viscosity η usually follows the well-known Einstein equation [33, 34]:

$$\eta = \eta_s \left(1 + 2.5\psi\right) \tag{1}$$

where η_s is the liquid viscosity, and ψ the particle volume fraction. The main assumptions of the Einstein equation are: the suspension contains uniform-size, zero-charge, hard spherical particles; there are no interactive forces between particles; and the suspension is dilute with a very-low-volume fraction ($\psi < 0.01$) and behaves as a Newtonian fluid. For non-dilute systems, the above equation has been modified to include the higher-order volume fraction terms to account for the influence of the surrounding particles, as given below [33, 34]:

$$\eta_r = \eta_s \left(1 + 2.5\psi + b\psi^2 + c\psi^3 \dots\right) \tag{2}$$

where η_r is relative viscosity.

Equations (1) and (2) are not applicable for highly interactive suspensions such as lithium-ion electrode slurries. For these suspensions, the viscosity is also dependent on the Brownian and hydrodynamic forces; particle sizes and shapes; maximum solids packing fraction; particle charges; particle/particle and particle/polymer interactions. Generally, they exhibit non-Newtonian behavior. In addition to the shear-dependent viscosity, the slurry may also exhibit time-dependency, yield stress, and viscoelasticity. Although there are numerous theoretical or semi-theoretical constitutive equations available in the literature to describe the non-linear behavior of interactive concentrated suspensions, these are either too complex for practical usage or limited to a specific system only. Hence, experimental measurements of rheological properties have been considered as a reliable tool for interactive complex slurry characterization. Interested readers can refer to textbooks devoted to this field [33, 34].

Most lithium-ion electrode slurries exhibit a shear-thinning behavior where the shear-viscosity reduces with increasing shear rate [16, 22, 35, 36]. This characteristic is more pronounced for higher-solid-content slurries [37]. At higher shear rates, the strong shear force tends to destroy the network structure, forcing the particles to realign themselves in a more orderly structure parallel to the shear field, causing the viscosity to decrease and reach a Newtonian plateau at high shear rates [16, 38]. Typical viscosity/shear rate behavior of lithium-ion electrode slurry showing different concentrations of conductive agent (carbon black) in binder solution (PVDF/NMP solvent) is shown in Figure 3.1. At low carbon black concentration (0.25wt%), the shear-viscosity is nearly constant, independent of shear rate, and the slurry can be considered a Newtonian slurry. The shear-thinning behavior becomes more obvious as the particle concentration increases. The increase in viscosity at very low shear rate is nearly a hundred-fold when the solid concentration increases from 0.25 to 4.0wt%, although not as much in the high shear rate region due to the effect of shear-thinning. This non-Newtonian characteristic is advantageous in electrode processing since a low viscosity slurry is required for good coating at high shear rate conditions, and a high viscosity medium can prevent or minimize particle sedimentation when the slurry is at rest or during drying [16].

3.3.2 Viscoelasticity

From a fluid mechanics point of view, viscoelasticity is important only in transient, accelerated or decelerated flows, and has no direct impact under steady flow conditions. However, this property is closely related to the

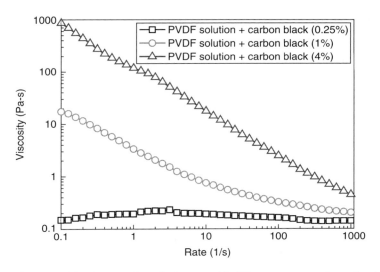

Figure 3.1 The viscosity data of mixtures with different concentrations of carbon black dispersed in the PVDF solution.

molecular or internal structure of the fluid or suspension. Rheological measurements of viscoelasticity are normally carried out under oscillatory shear mode where the storage and loss moduli, G′ and G″, are reported as a function of frequency or strain rate. Typically, many polymer melts, polymer solutions and solid/liquid suspensions exhibit a G′ greater than G″ at low frequencies until a critical frequency or strain rate is reached, when a crossover occurs beyond which G″ > G′. At low strain rate, the external force exerted on the material is insufficient to cause any disruption of the material's internal structure. The material behaves like an elastic solid. When the external force is strong enough to break the structure, the material flows like a viscous liquid. Solid/liquid suspensions exhibiting these characteristics are sometimes classified as "viscoplastic" materials. The corresponding stress at the critical strain has been shown to be related to the yield stress of the material [39, 40].

Figure 3.2 represents typical viscoelastic behavior of a lithium-ion electrode slurry measured at room temperature, where G′ and G″ are presented as a function of frequency [13–25]. Over the measured range of frequency (0.1 to 200 rad/s), the storage modulus is greater than the loss modulus, suggesting that the slurry is behaving like a solid material. Although there are no high frequency data reported due to instrument limitations, the figures appear to indicate that the transition occurs at a frequency of around 300 rad/s. The material will flow like a liquid at frequencies above this critical value.

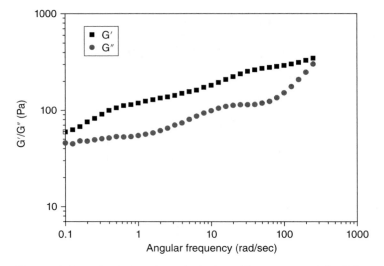

Figure 3.2 Viscoelastic data of a cathode slurry. The storage modulus (G') is greater than the loss modulus (G'') for the cathode slurry, which has 56wt% active material, 1.3wt% Super-P, 2.5wt% KS-6, 3.2wt% PVDF and 37wt% NMP.

3.3.3 Yield Stress

Yield stress is an inherent material property of many high solid content suspensions or slurries, as in the case of electrode slurries. When the applied stress is less than the yield stress, the slurry will not deform and will appear like a solid. However, when the yield stress is exceeded, the slurry will behave like a fluid, either a Newtonian or a pseudo-plastic fluid [39]. There are many methods in the literature for yield stress determination, mostly presented in either constant-stress steady or dynamic mode [39, 40].

The two simplest rheological models for yield stress of materials are:

Bingham model

$$\tau = \tau_0 + \mu\gamma \tag{3}$$

Herchel-Bulkley model

$$\tau(\gamma) = \tau_0 + \kappa\gamma^n \tag{4}$$

where τ is the shear force, γ is the shear rate, τ_0 is the yield stress, μ is the apparent viscosity, κ is the consistency coefficient, and n is the power law index. In the absence of yield stress, the Bingham model becomes the Newtonian model, and the Herchel-Bulkley model becomes the common pseudo-plastic power law model.

The presence of yield stress in electrode slurry may influence the material behavior during processing. In the preparation of slurry, a dead zone or a solid plug may be created inside the mixer. Hence, the mixer must be designed in such a way that during mixing, all components in the mixture are subjected to high shear everywhere within the mixer to ensure thorough mixing. The impact of yield stress on the coating process is relatively small since the shear stress or shear rate is sufficiently high during the usual coating operation of wet electrode film. However, during drying of wet film, the evaporation of solvent causes the solid content to increase, resulting in a higher yield stress. A higher yield stress is considered to be advantageous as it can hinder or minimize particle sedimentation under gravity.

3.4   Effects of Rheology on Electrode Processing

3.4.1   Composition of Electrode Slurry

Porcheret et al. [13] compared the rheological behavior of an LiFePO$_4$ cathode using organic and aqueous binder/solvent systems. The organic binder solution was PVDF in NMP, and the aqueous binder solution was either CMC or HPMC thickener in water. It was found that at the same viscosity level, the organic PVDF binder system exhibited a large yield stress, followed by the aqueous CMC system, while the HPMC system showed no yield stress phenomenon. Similar behavior was observed from the dynamic moduli measurements in which G′ was greater than G″ for the PVDF and CMC systems, but the opposite trend was seen for the HPMC system. This was consistent with the yield stress results where the former two slurries were more solid-like and the latter was more liquid-like when subjected to lower strain motion. As indicated by the simulation results presented by Zhu et al. [30], particle aggregation and distribution in the electrode slurry was strongly dependent on the quantities and sizes of active material and conductive agent added. Ligneel et al. [14] also found that increasing solid content in the slurry would increase the yield stress, which would affect solvent evaporation and binder migration during the drying process.

Cho et al. [22] studied the effect of slurry storage time by comparing the thixotropic (time-dependent) characteristics of two cathode slurries, one resting for a day and the other for seven days after formulation. It was found that both steady shear and dynamic viscosities were higher for the one-day sample than for the seven-day sample. The slurry impedance was also found to increase continuously over a three-day period.

In order to overcome the settling issue of the active material in the aqueous system, Bitsch et al. [24] added octanol to the mixture to increase the slurry viscosity at stationary state. However, with the effect of shear-thinning, it had no direct impact on the coating process as the slurry was moving at relatively high shear rate conditions.

Table 3.2 is a summary of the rheological properties of lithium-ion electrodes reported by various researchers.

3.4.2 Electrode Slurry Preparation

3.4.2.1 Mixing Methods

The homogeneity of electrode slurry is critical in the production of wet electrode film through coating and drying processes, and overall battery performance. In the mixing process, the sequence of adding various electrode components and the mixing method affects the slurry rheology and the degree of particle dispersion in the mixture. There are many mixing sequences proposed in the literature [15–17, 19, 20, 26–30, 32]. In general, most suggested mixing methods involved three steps. The initial step was the mixing of active material and conductive agent, both in powder form, to establish a good conductive network [33, 34]. This was followed by blending in the pre-mixed binder solution to provide good adhesion between particle/particle and particle/current collector. Finally, solvent and/or dispersant was added to the mixture to adjust the slurry viscosity. During the addition of electrode ingredients, the mixture must be continuously mixed to ensure good dispersion.

In order to improve the dispersion of the slurry, the addition of solvent in multiple stages was recommended [15–17, 20]. Initially, small quantities of solvent were added sequentially to the solid mixture until it reached the maximum packing fraction. Prior to the coating of the slurry, an additional amount of solvent and a small quantity of dispersant (surfactant) were added to lower the slurry viscosity to a level suitable for coating of wet film.

A brief review of some mixing methods [15–17, 19, 20, 26–30, 32], the choice of dispersants [13, 21, 23, 24, 31], and their effects on the battery manufacturing processes and performance are briefly discussed here.

Kim *et al.* [15] compared the dispersion of the slurry using four different electrode component mixing sequences and the effect on the battery performance. Firstly, the conductive additive was mixed thoroughly with solvent and binder, followed by the addition of the active material. Secondly, the order of adding the solid ingredients, active material and conductive agent, was reversed. Thirdly, all ingredients were mixed together at once. The final method was to pre-mix the two solid components in dry state, blend in the binder and solvent in sequence, and then vigorously stir the mixture until complete dispersion was achieved. The four slurries exhibited different steady shear-viscosity behavior, with the one produced by the last method yielding the lowest viscosity level under the same shear conditions. This indicated that the dispersion of this slurry was better than that of the other three slurries. Subsequent battery test on a battery plate produced from this

Table 3.2 Rheological analyses of various lithium-ion battery slurries.

Active material	Binder	Conductive agent	Solvent	Additive	Rheological measurement	Ref
$LiFePO_4$	PVA-PEG	Super-P (Carbon black)	Deionized water	HPMC, CMC (thickener) Triton X-100 (surfactant)	• Elastic modulus vs strain • Viscosity vs shear rate • Shear stress vs shear rate • Storage/loss modulus vs frequency	[13]
$Li_{1.1}V_3O_8$	PMMA	Super-P (Carbon black)	THF	EC-PC	• Shear stress vs shear rate	[14]
$LiCoO_2$	PVDF	Graphite	NMP	Non	• Viscosity vs time	[15]
$LiCoO_2$	PVDF	Denka black	NMP	Non	• Viscosity vs shear rate • Shear stress vs shear rate • Storage/loss modulus vs frequency	[16]
$LiFePO_4$	SBR	Super-P (Carbon black), KS-6 (graphite)	Deionized water	SCMC	• Viscosity vs shear rate • Storage/loss modulus vs frequency	[17]
$LNi_{1/3}Co_{1/3}Mn_{1/3}O_2$	PVDF	Acetylene black	NMP	Non	• Viscosity vs shear rate	[18]
LMS/LNCA $Li(Li-Mn-Al)_2O_4/$ $LiNi_{0.8}Co_{0.15}Al_{0.05}O_2$	PVDF	Super-P (Carbon black)	NMP	Non	• Viscosity vs time	[19]
$LNi_4Co_2Mn_4O_2$	PVDF	Super-P (Carbon black), KS-6 (graphite)	NMP	Non	• Viscosity vs shear rate • Shear stress vs shear rate	[20]
$LiFePO_4$	CMC	Carbon black	Deionized water	PAA	• Viscosity vs shear rate	[21]

(Continued)

Table 3.2 (Continued)

Active material	Binder	Conductive agent	Solvent	Additive	Rheological measurement	Ref
LiCoO₂	PVDF	Super-P (Carbon black)	NMP	Non	• Viscosity vs shear rate • Shear stress vs shear rate • Storage/loss modulus vs frequency	[22]
Si	CMC SBR	Super-P (Carbon black)	Buffer solution	PAMA	• Viscosity vs shear rate • Viscosity vs time	[23]
Graphite	SBR	Carbon black	Deionized water	CMC Octanol	• Viscosity vs shear rate • Storage/loss modulus vs frequency	[24]
Graphite	PVDF	Carbon black, Graphite	NMP	Non	• Viscosity vs shear rate • Storage/loss modulus vs frequency	[25]

slurry also resulted in a more stable battery discharge capacity and cycle-life characteristics.

Lee *et al.* [16] compared the rheology of two cathode slurries produced by different mixing methods. The first method was to thoroughly mix the liquid phase (PVDF binder and NMP solvent) and the solid phase (active material $LiCoO_2$ and conductive carbon powder) separately, and then combine the two phases. The second method was to prepare a solution of PVDF containing only 2/5 of the solvent (NMP), then mix this with the solid mixture, followed by the addition of the remaining solvent in equal portions in three successive intervals with continued stirring. The slurry using the multi-stage addition of solvent exhibited a lower steady shear-viscosity, and higher dynamic loss modulus G″ than storage modulus G′ The battery made from this cathode slurry also had a better C-rate discharge characteristic.

Li *et al.* [17] studied the battery quality produced from aqueous electrodes with two mixing methods. The binder used was SBR (styrene-butadiene rubber) and the thickening solution was SCMC (sodium carboxymethyl cellulose) in water. The first method was to continuously mix together the thickener with the active material ($LiFePO_4$) and the conductive additive (KS-6 graphite and carbon black) for one day. The binder (SBR) was then added and stirred for an additional three days. The second method was to pre-mix the binder and thickeners first, then mix with the active material and conductive additive, and continuously stir the mixture for four days. A lower slurry viscosity was obtained using the first mixing method, indicating better particle dispersion in the slurry. The SEM analysis also revealed significant particle aggregation on the surface of the electrode produced from the slurry mixed by the second method. Subsequent battery performance test also indicated that the battery fabricated from slurry using the multi-step mixing method (the first method) had a better high C-rate and good charge/discharge cycle characteristics.

3.4.2.2 Mixing Devices

The ball mill mixer is one of the most commonly used devices for mixing electrode slurries. However, the energy consumption for this type of mixer is large and it is impractical for large-scale production, although it has been industrially accepted as a reference mixer for electrode slurry preparation [37]. Other types of mixers such as planetary mixers, high speed mixers, and homogenizers have been used in industry for electrode slurry mixing [41, 42]. A necessary condition for a good mixer is to ensure that all particles everywhere in the container must be in constant motion during mixing. This can only be achieved with a mixer that can generate a uniform flow field, to the greatest extent possible, in all directions. Aiming to satisfy this condition in the shortest possible mixing time, Liu *et al.* [20] recently developed a new laboratory-scale, 3-dimensional cylindrical mixer, which enabled

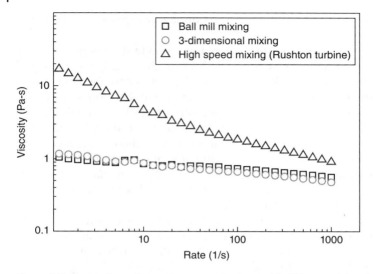

Figure 3.3 Comparison of viscosity curves obtained with different mixing devices.

simultaneous mixing in r, θ, and z directions, and was successfully used for mixing lithium-ion cathode slurries.

Figure 3.3 shows comparison of the shear-viscosity of three cathode slurries mixed in different devices. A significant shear-thinning behavior was observed for the slurry mixed in a Rushton turbine mixer for two hours and 40 minutes, as indicated by the upper curve. The viscosity continued to drop with increasing shear rate over a long period of time, implying that complete mixing had not been achieved. The other two viscosity curves, obtained from the ball mill and the 3D mixers, appeared to approach a relatively constant Newtonian viscosity at a lower shear rate, suggesting complete particle dispersion. However, the mixing times for these two mixers were significantly different. It took three days of continuous mixing in the ball mill as compared to an hour in the 3D mixer to achieve similar results. Electrical performance test of the battery made with the slurry prepared in multiple stages in the 3D mixer also yielded good battery characteristics and cycle life.

In order to further facilitate good dispersion of particles during mixing, a small amount of dispersant may be added to the mixture. Zhang *et al.* [31] added a small amount of surfactant (Triton X-100) into an LiFePO$_4$ electrode, and compared the electrochemical properties of this electrode with those of an electrode without added surfactant. The results showed that there was a substantial improvement in electrochemical properties for battery capacity and cycle life of the electrode with added surfactant.

3.4.3 Electrode Coating

In the coating and printing industries, solutions with a wide range of viscosity are encountered. For example, the viscosity could be as low as 1 mPa-s in gravure and injection printings and in spray coating, and as high as 50,000 mPa-s in offset and screen printings. Depending on the application, the slot-die coating has been shown to be a more versatile coating process. It could handle coating liquid/suspension with a wide range of viscosity. In the production of battery electrode film using the slot-die coating process, the shear-viscosity has a strong influence on the film stability, coating uniformity and operating coating window [43–46]. Usually when a solid–liquid slurry contains polymer, the viscosity and surface tension would be higher due to the particle/polymer interaction, which would influence the operating window in the slot-die coating [47]. In addition to shear-thinning, yield stress could also be present for coating of the high solid content electrode slurry. This would affect the velocity distribution inside the slot-die, and might cause an uneven film thickness or film splitting [48]. Multi-layered simultaneous slot-die coating was recently accepted as a new technology for multifunctional electrode production. In simultaneous two-layer coating of electrodes, it was necessary to have two slurries with similar rheological properties to ensure uniform coating. The thickness ratio of the upper/lower layer was an important parameter affecting the minimum film thickness, and was primarily controlled by the viscosities of the two coating slurries [49–55].

3.4.4 Electrode Drying

Following the production of wet electrode film by coating or other methods, the film must be dried, usually in a dryer, for solvent evaporation and curing. The final internal structure of electrode would be established after drying. Although there is no external shear or extensional flow directly applied to the wet film as it moves through the dryer, rheology could still play an important role in determining the internal structure of the film. The solid content of the wet film, particle distribution and conduction network configuration would change during drying due to solvent evaporation together with the migration of binder toward the exposed heated surface [56–62]. Particle sedimentation under gravity could also occur. The presence of yield stress would prevent this phenomenon from happening or minimize it. The viscosity of the liquid in the porous medium structure would affect the movement of solvent and binder due to its effects on mass diffusion and capillary driving forces. The change in the electrode internal structure could be detected from the behavior of the dynamic moduli obtained [63].

Finally, the rheological behavior of electrode slurry would be sensitive to temperature changes. In industrial applications, the slurry temperature during coating is normally controlled within the range 35–65°C, and drying at around

80–150° C. These temperature ranges were shown to make coating and drying of electrode slurries more efficient due to their effects on the slurry rheology [64].

3.5 Conclusion

Lithium-ion battery slurry is a very complex mixture consisting of solvent, binder, active material and conductive additive. It generally behaves as a non-Newtonian, viscoplastic material. It exhibits various rheological characteristics including shear-thinning viscosity, viscoelasticity, yield stress and time-dependency. The electrochemical performance of the battery depends strongly on the manufacturing process of the electrodes. Understanding slurry rheology is essential as it influences the entire electrode manufacturing process, from the composition of the electrode components, including chemicals used, particle size and shape, to the preparation of the slurry, involving the mixing procedure and selection of appropriate mixers, to the production of wet film by coating, and finally the drying and curing of wet electrodes.

List of Symbols and Abbreviations

CMC, carboxymethyl cellulose
EC, ethylene carbonate
HPMC, hydroxypropyl methylcellulose
NMP, N-methyl-2-pyrrolidone
PAA, poly acrylic acid
PAMA, poly (acrylic-co-maleic) acid
PC, propylene carbonate
PVA-PEG, polyvinyl alcohol-polyethylene glycol
SBR, styrene-butadiene rubber
SCMC, sodium carboxymethyl cellulose
THF, Tetrahydrofuran

References

1 Manthiram, A., Nazari, C.A. (2009) *Lithium Batteries Science and Technology*, Springer, 1–37.
2 Tarascon, J.M., Armand, M. (2001) *Nature* **414**, 359–367.
3 Whittingham, M.S. (2000) *Solid State Ionics* **134**, 169–178.
4 Matsuki, K., Ozawa, K. (2009) *Lithium Ion Rechargeable Batteries*, Wiley-VCH, Weinheim, 1–9.
5 Wagner, F.T., Lakshmanan, B., Mathias, M.F. (2010) *J. Phys. Chem. Lett.* **1**, 2204–2219.

6 Horiba, T. (2014) *Proceedings of the IEEE* **102**, 939–950.
7 Choi, H.S., Park, C.R. (2010) Towards high performance anodes with fast charge/discharge rate for LIB based electrical vehicles. In C.R. Park (ed) *Lithium-ion Batteries*, InTech, 1–24.
8 Cheon, S.E., Kwon, C.W., Kim, D.B., Hong, S.J., Kim, H.T., Kim, S. W. (2000) *Electrochim. Acta* **46**, 599–605.
9 Yoshio, M., Brodd, R.J., Kozawa, A. (2009) *Lithium-ion Batteries: Science and Technologies*, Springer, 183–185.
10 Fergus, J.W. (2010) *J. Power Sources* **195**, 939–954.
11 Zheng, H., Yang, R., Liu, G., Song, X., Battaglia, V.S. (2012) *J. Phys. Chem.* **C 116**, 4875–4882.
12 Li, J., Daniel, C., Wood, D. (2011) *J. Power Sources* **196**, 2452–2460.
13 Porcher, W., Lestries, B., Jouanneau, S., Guyomard, D. (2009) *J. Electrochem. Soc.* **156**, A133–A144.
14 Ligneel, E., Lestriez, B., Hudhomme, A., Guyomard, D. (2007) *J. Electrochem. Soc.* **154**, A235–A241.
15 Kim, K.M., Jeon, W.S., Chung, I.J., Chang, S. H. (1999) *J. Power Sources* **83**, 108–113.
16 Lee, G.W., Ryu, J.H., Han, W., Ahn, K.H., Oh, S.M. (2010) *J. Power Sources* **195**, 6049–6054.
17 Li, C.C., Lin, Y.S. (2012) *J. Power Sources* **220**, 413–421.
18 Zheng, H., Tan, L., Liu, G., Song, X.G., Battaglia, V.S. (2012) *J. Power Sources* **208**, 52–57.
19 Haselrieder, W., Ivanov, S., Tran, H.Y., Theil, S., Froböse, L., Westphal, B. *et al.* (2014) *Progress in Solid State Chemistry* **42**, 157–174.
20 Liu, D., Chen, L.C., Liu, T.J., Fan, T., Tsou, E.Y., Tiu, C. (2014) *Adv. Chem. Eng. Sci.* **4**, 515–528.
21 Lee, J.H., Kim, J.S., Kim, Y.C., Zang, D.S., Paik, U. (2008) *Ultramicroscopy* **108**, 1256–1259.
22 Cho, K.Y., Kwon, Y.I., Youn, J.R., Song, Y.S. (2013) *Analyst* **138**, 2044–2050.
23 Nguyen, B.P.N., Chazelle, S., Cerbelaud, M., Porcher, W., Lestriez, B. (2014) *J. Power Sources* **262**, 112–122.
24 Bitsch, B., Dittmann, J., Schmitt, M., Scharfer, P., Schabel, W., Willenbacher, N. (2014) *J. Power Sources* **265**, 81–90.
25 Yoo, M., Frank, C.W., Mori, S. (2003) *Chem. Mater.* **15**, 850–861.
26 Kim, J., Eom, M., Noh, S., Shin, D. (2013) *J. Power Sources* **244**, 476–481.
27 Ponrouch, A., Palacín, M.R. (2011) *J. Power Sources* **196**, 9682–9688.
28 Zheng, H., Zhang, L., Liu, G., Song, X., Battaglia, V.S. (2012) *J. Power Sources* **217**, 530–537.
29 Nakajima, H., Kitahara, T., Higashinaka, Y., Nagata, Y. (2015) *ECS Transaction* **64(22)**, 87–95.
30 Zhu, M., Park, J., Sastry, A.M. (2011) *J. Electrochem. Soc.* **158**, A1155–A1159.
31 Zhang, W., He, X., Pu, W., Li, J., Wan, C. (2011) *Ionics* **17**, 473–477.

32 Liu, D., Chen, L.C., Liu, T.J., Tiu, C., Fan, T., Tsou, E.Y. (2013) *The 5*th *Asian Coating Workshop.*

33 Malkin, A.Y., Isayev, A.I. (2006) *Rheology: Concepts, Methods, and Applications*, ChemTec Publishing, Toronto, 146–147.

34 Macosko, C.W. (1994) *Rheology: Principles, Measurements, and Applications*, Wiley-VCH, New York.

35 Despotopoulou, M., Burchill, M.T. (2002) *Prog. Org. Coat.* **45**, 119–126.

36 Usui, H., Kishimoto, K., Suzuki, H. (2001) *Chem. Eng. Sci.* **56**, 2979–2989.

37 Stickel, J.J., Powell, R.L. (2005) *Annu. Rev. Fluid Mech.* **37**, 129–149.

38 Carreau, P.J., Dekee, D., Chhabra, R.P. (1997) *Rheology of Polymeric Systems: Principles and Applications*, Hanser, Munich.

39 Uhlherr, P.H.T., Guo, J., Tiu, C., Zhang, X.M., Zhou, J.Z.Q., Fang, T.N. (2005) *J. Non-Newt. Fluid Mech.* **125**, 101–119.

40 Tiu, C., Guo, J., Uhlherr, P.H.T. (2006) *J. Ind. Eng. Chem.* **12**, 5 653–662.

41 Tanguy, P.A., Thibault, F., Dubois, C., Ait-Kadi, A. (1999) *Chem. Eng. Res. Des.* **77**, 318–324.

42 Zhou, G., Tanguy, P.A., Dubois, C. (2000) *Chem. Eng. Res. Des.* **78**, 445–453.

43 Gutoff, E.B., Cohen, E.D., Kheboian, G.I. (2006) *Coating and Drying Defects: Troubleshooting Operating Problems*, Wiley-Interscience, Hoboken, NJ.

44 Ruschak, K.J. (1976) *Chem. Eng. Sci.* **31**, 1057–1060.

45 Higgins, B.G., Scriven, L.E. (1980) *Chem. Eng. Sci.* **35**, 673–682.

46 Lee, K.Y., Liu, L.D., Liu, T.J. (1992) *Chem. Eng. Sci.* **47**, 1703–1713.

47 Chu, W.B., Yang, J.W., Wang, Y.C., Liu, T.J., Tiu, C., Guo, J. (2006) *J. Colloid Interface Sci.* **297**, 215–225.

48 Nagashima, M., Hasegawa, T., Narumi, T. (2006) *J. Soc. Rheol. Japan* **34**, 213–221.

49 Yu, W.J., Liu, T.J. (1995) *Chem. Eng. Sci.* **50**, 917–920.

50 Lu, S.Y., Lin, Y.P., Liu, T.J. (2001) *Polym. Eng. Sci.* **41(10)**, 1823–1820.

51 Lin, Y.N., Huang, S.Y., Liu, T.J. (2005) *Polym. Eng. Sci.* **45**, 1590–1599.

52 Schmitt, M., Raupp, S., Wagner, D., Scharfer, P., Schabel, W. (2015) *J. Coat. Technol. Res.* **5**, 877–887.

53 Chen, L.C., Liu, D., Liu, T.J., Tiu, C., Yang, C.R., Chu, W.B. (2016) *J. Energy Storage* **5**, 156–162.

54 Chen, L.C., Liu, D., Liu, T.J., Tiu, C., Yang, C.R., Chu, W.B. (2015) *The 7*th *Asian Coating Workshop.*

55 Liu, D., Chen, L.C., Liu, T.J., Tiu, C., Chu, W.B. (2015) *The 7*th *Asian Coating Workshop.*

56 Bernada, P., Bruneau, D. (1996) *TAPPI Journal* **79**, 130–143.

57 Zhen, Z., Wang, Z. (2013) *Adv. Mater. Research* **790**, 45–48.

58 Li, Y., Gu, W., He, B. (2014) *Adv. Mater. Research*, **881–883**, 1460–1463.

59 Vanderhof, J., Bradford, E. (1990) *TAPPI Coating Conference Proceedings*, Atlanta, USA, 173–177.

60 Ranger, A.E., Flnstp, C. (1994) *Paper Tech.* **35**, 40–46.

61 Westphal, B.G., Bockholt, H., Günther, T., Haselrieder, W., Kwade, A. (2015) *ECS Transaction* **64(22)**, 57–68.

62 Jaiser, S., Müller, M., Baunach, M., Bauer, W., Scharfer, P., Schabel, W. (2016) *J. Power Sources* **318**, 210–219.

63 Komoda, Y. (2013) *The 5th Asian Coating Workshop*, Seoul, Korea.

64 Wakamatsu, H., Natsume, T. (2013) U.S. Patent 0056092.

4

Polymer Electrolytes for Printed Batteries

Ela Strauss[1], Svetlana Menkin[2] and Diana Golodnitsky[2,3]

[1] Ministry of Science, Space and Technology, Jerusalem, Israel
[2] School of Chemistry, Tel Aviv University, Israel
[3] School of Applied Materials, Tel Aviv University, Israel

The huge growth of the printed-microelectronics industry has resulted in many efforts being devoted to the development of printable power sources. Printing of power sources has become a necessity for portable autonomous devices, RFID cards, wearable electronics and IoT (Internet of Things) applications [1]. Printing or painting enables the design and integration of batteries on various substrates and geometries including curved and flexible surfaces with confined spaces. Several attempts have been made to produce primary and secondary thin-film batteries utilizing printing techniques exclusively. These technologies are still at an early stage, and most currently printed batteries exploit printed electrodes sandwiched with self-standing polymer membranes, produced by conventional extrusion, papermaking techniques, stretching, evaporation, and sintering methods followed by soaking in aqueous or nonaqueous liquid electrolytes. Printing of solid polymer and gel electrolytes (GPE) remains a bottleneck in the all-printed batteries and the literature is scarce. In this short communication, an attempt has been made to address the issue of polymer electrolyte printing, to compare the electrochemical performance of conventional bulk batteries and printed batteries with polymer electrolytes and to propose avenues of investigation and development in this rapidly growing field.

4.1 Electrolytes for Conventional Batteries

Conventional batteries are fabricated by superposition of positive and negative electrodes coated on current collectors with an intermediate layer of a separator or polymer electrolyte. The separators (ion-conducting membranes) are

Printed Batteries: Materials, Technologies and Applications, First Edition.
Edited by Senentxu Lanceros-Méndez and Carlos Miguel Costa.
© 2018 John Wiley & Sons Ltd. Published 2018 by John Wiley & Sons Ltd.

produced as free-standing films. Ion-conducting membranes play a key role in electrochemical energy-storage devices. Good battery performance necessitates sufficiently high ionic conductivity in the bulk electrolyte, as well as a supply of cations to both cathode/electrolyte and anode/electrolyte interfaces to sustain fast charge–discharge processes [2]. Ion transport can take place through intrinsic conduction paths of the membrane or via impregnated liquid electrolyte. Separator/electrolyte compositions must be tuned to conduct ions easily, while simultaneously forming safe, impenetrable and electronically insulating barrier layers. In order to minimize the internal impedance of the battery, the separator should follow the contours of the relatively rough surface of the electrode materials. In addition, in rechargeable batteries the membranes/electrolytes must have the high mechanical properties needed to withstand the changes in the electrode volumes during the operation of the cell.

For aqueous-electrolyte batteries (nickel–cadmium, nickel–metal hydride, manganese-dioxide, zinc-silver oxide, zinc-air, zinc-mercuric oxide), with impregnated NaOH- or KOH-based alkaline electrolytes, the typically used materials for separators are: polyvinylchloride (PVC), polyamide (nylon), polypropylene (PP), regenerated cellulose ('cellophane'), irradiated polyethylene, microporous polypropylene pretreated to increase hydrophilicity. For lead-acid or carbon-lead-acid cells, the materials for separators are cellulose, polyvinyl chloride, organic rubber, polyolefin, glass, synthetic pulp and polyethylene-silica composite.

4.1.1 Polymer/Gel Electrolytes for Aqueous Batteries

In the construction of aqueous-electrolyte-based rechargeable batteries, there is a major advantage in using a solid or gel electrolyte instead of a liquid electrolyte, in terms of reliability, safety, easy design and processability. However, there have been only a few reports of a solid or gel electrolyte with high ionic conductivity for use in aqueous-electrolyte-based batteries. For example, in the zinc/carbon battery (Leclanché), a paper separator coated with gelling agent such as methylcellulose is used, which results in a gelled alkaline electrolyte absorbed into the paper separator.

The valve-regulated lead-acid battery (VRLA) is often referred to as a sealed and/or maintenance-free lead-acid battery. It differs from the conventional flooded lead-acid battery in containing only a limited amount of electrolyte ("starved" electrolyte) absorbed in a separator or immobilized in a gel. The electrolyte is commonly immobilized by two methods. In the first, a highly porous absorbing mat, fabricated from microglass fibers, is partially filled with electrolyte, and acts as the separator/electrolyte reservoir. This is referred as "absorbed electrolyte". In the second, referred to as the gelled-electrolyte method, fumed silica is added to aqueous acid electrolyte, causing it to harden into a gel. The immobilization of the electrolyte allows the battery to operate in different orientations without spillage [3].

4.1.2   Electrolytes for Lithium-ion Batteries

The high energy and power requirements of printed batteries suitable for diverse applications is one reason why a lithium-ion technology should be preferred over other available types of battery. The electrolytes for lithium or lithium-ion batteries are based on nonvolatile, thermally and electrochemically stable aprotic solvents. The conductivity of the electrolyte should be at least 3 mS·cm^{-1} to make it practical. The most widely used salts are lithium hexafluorophosphate (LiPF$_6$) and lithium tetrafluoroborate (LiBF$_4$) [4]. In addition, a wide variety of alternative solvents as well as additives have been proposed, mainly aiming to improve the safety and compatibility of the electrolytes with high-voltage cathodes. These include groups of co-solvents with fluorinated alkyl carbonates, which enhance both electrode passivation and thermal stability, and non-flammable components such as organophosphorus compounds [5].

Currently developed high-voltage (4.3 V) cathode materials are beyond the voltage window of typical carbonate-based electrolytes; thus the organic electrolyte undergoes apparently continuous oxidative decomposition during cycling and forms non-passivating solid-electrolyte-interphase (SEI) films on electrodes. Some examples of additives for electrochemically stable high-voltage electrolytes are biphenyl and other aromatic molecules such as propane-sultone, butanesultone, etc.

In recent years, ionic liquids (ILs) as electrolyte additives or solvents, gel, hybrid, solid-polymer and ceramic electrolytes have begun to progressively replace electrolytes based on organic solvents, mainly with the purpose of improving the safety and compatibility of the electrolytes with high-voltage cathodes. ILs are characterized by unique properties, such as non-flammability; negligible vapor pressure; remarkable ionic conductivity; high thermal, chemical, and electrochemical stabilities; low heat capacity; and ability to dissolve inorganic (including lithium salts), organic, and polymeric materials. Viable lithium-battery ILs are formed by alkyl imidazolium, saturated alkyl quaternary ammonium cyclic cations (pyrrolidinium, piperidinium) or acyclic (tetra alkyl ammonium) cations, in combination with hydrophobic perfluoro-alkyl sulfonyl imide anions [4].

Gel electrolytes for lithium-ion batteries are composed of a polymer host, solvent and lithium salt. Interacting strongly in liquid organic solutions, polymers generally form chemically stable gel electrolytes [4]. Poly(vinylidenefluoride) (PVDF)-based polymer electrolytes are the most common lithium-ion gel electrolytes. Because of the strongly electron-withdrawing functional group (-C-F), PVDF can assist in greater ionization of lithium salts, thus providing a high concentration of charge carriers. Different approaches with the aim of decreasing the crystallinity of the PVDF matrix, and improving the mechanical properties and interfacial stability include: (1) use of branched PVDF-HFP or PVDF–TrFE copolymers rather than linear PVDF monomers, (2) incorporation of nanometric fillers such as $BaTiO_3$, Al_2O_3, SiO_2, and TiO_2 and (3) addition of different polymeric components to obtain blend-polymer hosts [4]. The first commercial GPE Li-ion cell was based on PVDF-hexafluoropropylene (HFP) copolymer electrolyte [6]. The conductivities of blend-polymer systems are between 0.98 and 4.36 $mS \cdot cm^{-1}$ at 20 °C. The addition of poly(methyl methacrylate) (PMMA) decreases the crystallinity and increases the pore size, porosity, electrolyte uptake and ionic conductivity of PMMA-PVDF gel-polymer electrolyte [7]. Prasanth *et al.* [8] reported a polymer blend of PAN/PMMA/ polystyrene (PS) electrolyte for lithium-ion batteries which have thermal stability up to 295 ± 5 °C and ionic conductivity of about 4 $mS \cdot cm^{-1}$.

Polymers, such as poly(ethylene oxide) (PEO), which strongly interact with solvent, tend to form very stable gels with conventional organic electrolytes, but these are characterized by very poor mechanical properties. Gel-polymer electrolytes based on high-molecular-weight PEO are sticky, highly viscous fluids with ionic conductivity above 0.1 $mS \cdot cm^{-1}$. In order to obtain the best compromise between high conductivity, homogeneity and dimensional stability, Choi *et al.* [9] proposed a hybrid solid electrolyte consisting of PEO, $LiClO_4$, a mixture of ethylene carbonate (EC), γ-butyrolactone (BL) and poly(acrylonitrile) (PAN). The highest room-temperature conductivity of 2 $mS \cdot cm^{-1}$ is found for a hybrid film of $31PEO-9LiClO_4 -50EC/BL-10PAN$. This film has a conductivity similar to that of PAN-based gel electrolytes, but with better dimensional stability.

Sony's laminate-type lithium-ion batteries exploit hybrid-polymer-gel technology. The liquid organic electrolyte is locked within the polymer and kept in a semi-solid state. The replacement of the liquid electrolyte by a semi-solid electrolyte results in more flexible battery shape, prolonged cycle life (1000 cycles with 90% high-power capability after a year), higher energy density due to tighter package, and greater safety, less swelling and no leakage as compared to traditional lithium-ion batteries [10].

Solid inorganic ceramic and polymer electrolytes (SPEs) have been intensively studied as safe electrolytes for application in solvent-free lithium-ion rechargeable batteries. This is mainly motivated by their advantages, such as flame-resistance and flexibility. Using solid instead of liquid electrolytes in batteries simplifies encapsulation of battery components and hence increases shelf life and the range of operating temperatures [11]. Inorganic ceramics and organic polymers used for solid electrolytes in lithium-ion batteries differ in their mechanical properties. The high elastic moduli of ceramics make them more suitable for rigid-battery designs as in, for example, thin-film-based devices. Conversely, the low elastic moduli of polymers are useful for flexible battery designs. Conventional SPEs are prepared by dissolving a lithium salt in a polymer matrix, which, in some cases, also contains plasticizers and ceramic fillers. The polymer matrix must contain a Lewis base (e.g. one containing an ethylene oxide unit, $-OCH_2CH_2-$) to solvate the lithium salt. Over the past

30 years, great efforts have been made to develop polymer matrices with various structures for improving ambient-temperature ionic conductivity; however, up to now, progress on all-solid-state polymer electrolyte systems has been relatively slow. In fact, the lithium-ion polymer battery (abbreviated variously as LiPo, LIP, Li-poly, among others), is a rechargeable battery of lithium-ion technology in a flexible pouch format. The electrolyte used in this type of battery is of the gel type. Electrochemical cells with solid-polymer electrolytes have not reached full commercialization, and are still a topic of research [12].

Solid ceramic electrolytes have been employed primarily in thin-film batteries. One of the examples is amorphous material called LiPON (lithium phosphorous oxynitride), which is a nonstoichiometric material of lithium, phosphorus, nitrogen, and oxygen. This electrolyte has higher voltage stability than do organic polymer electrolytes, so the LiPON battery can accommodate higher voltage positive materials such as lithiated cobalt oxide or manganese oxide. Recently, Kato *et al.* [13] discovered that lithium superionic conductors $Li_{9.54}Si_{1.74}P_{1.44}S_{11.7}Cl_{0.3}$ and $Li_{9.6}P_3S_{12}$ showed the highest ionic conductivity (25 mS$\cdot$cm^{-1}) reported for lithium-ion, and high electrochemical stability versus lithium metal. Typical thin-film ceramic-based electrolyte materials used in current commercial variations of thin-film all-solid-state batteries are deposited in vacuum chambers by RF and DC magnetron sputtering and by thermal evaporation. In addition, many publications report exploring a variety of physical (PVD) and chemical vapor-deposition (CVD) processes, such as pulsed-laser deposition, electron cyclotron resonance sputtering, and aerosol spray coating [14]. It should be mentioned that ceramic films are rigid and brittle and create poor electrode/electrolyte interfacial contacts, leading to high battery impedance. The methods of their fabrication are relatively slow and require expensive machinery [3]. The THINERGY® MEC220 battery, produced by Infinite Power Solutions®, is a solid-state, rechargeable, thin-film Micro-Energy Cell (MEC). The active materials in the device include a lithium cobalt oxide ($LiCoO_2$) cathode and a lithium-metal anode; LiPON is used as a solid electrolyte [15]. While this battery is rechargeable and safe, its volumetric energy density (about 20Whl^{-1}) and capacity per footprint (about 0.08mAh$\cdot$cm$^{-2)}$ are lower by one order of magnitude than those of commercial lithium-ion batteries with liquid organic electrolytes.

4.2   Electrolytes for Printed Batteries

Currently, most printed batteries employ printed current collectors and printed electrodes sandwiching different membranes impregnated with liquid electrolytes [1, 16–19]. As in the conventional battery, high ionic conductivity via liquid-soaked membrane and compatibility with electrode materials are the essential properties for enabling high performance of the printed battery.

The membrane/electrolyte should be mechanically strong and rigid enough to sustain any mechanical stress, shock or tear. In addition, it should have sufficient flexibility to be able to bend or fold and conformally cover the electrode surface in order to avoid short-circuiting between layers during application of the next layer or during operation of the battery. The membrane should also display a certain degree of porosity and be able to take up appropriate amounts of liquid electrolyte. The membrane/electrolyte must be adherent to the electrode in order to avoid delamination and maintain the structural integrity of the printed battery. Very few publications appear in this field, with quasi-solid polymer gel or hybrid (swollen) membranes the most studied. The approaches for preparation of the separator/electrolyte layer by printing methods are still under investigation and yet again very few procedures are available in the literature. Therefore, the major challenge in the fabrication of a completely printed battery is the development of electrolyte inks.

Up to now, various coating procedures for electrode materials have been developed, most of them relying on solution-based processing. Direct, inkjet, screen or laser printing and stencil brush or spray painting are the most representative of this class. Each technique was found to be suitable for a different application, classification being primarily according to the process scale at which it can be implemented [1]. For example, laser writing is suitable for microscale batteries, printing for centimeter-scale, and painting for meter-scale power sources. All these techniques are considered to be applicable for the fabrication of electrolytes and are exemplified below.

4.2.1 Screen-printed Electrolytes

Screen-printing has been applied mainly to the fabrication of positive and negative electrodes for LIBs and supercapacitors. The technique is suitable for thin-film as well as for thick-film deposition and the lateral resolution is limited by the pattern mask. Doctor-blade (DB) and bar coating are techniques derived from the screen-printing approach [1]. DB coating is the most commonly used printing technique for depositing battery slurries in the large-scale manufacturing of conventional batteries as well [20]. It is worth mentioning that the doctor-blade coating is a blanket-coating process, such that it cannot be used to pattern inks over substrate as is possible with direct-writing methods such as inkjet printing or laser printing.

It is obvious that in order to fabricate a cell using printing technologies, all layer materials must be available as printing inks (pastes). The rheology requirements of screen-printing inks (viscosity, shear stress and shear rate) are similar to those of the standard battery. The inks for the electrodes are composed of active-material powder (anodic or cathodic), a polymeric binder, ceramic powder (for enhancing the mechanical strength of the film), electrically conductive materials (usually carbon-based) and an ionic conductive salt.

The ink formulation, depending on the required film thickness for each layer (anode, cathode), is chosen or developed on the basis of the requirements of the amount of deposited active material (required capacity, energy density, power density), electrical properties, electrode stability and electrolyte penetration. The viscosity of the ink is one of the parameters that control the film thickness. High ink viscosity and low vapor pressure of the solvent will result in thick electrodes. All the above-mentioned requirements are applicable to printed polymer electrolytes, in which the major component is dielectric polymer material.

Wright *et al.* [21] created a direct layer-by-layer printed zinc-based secondary battery with an ionic-liquid-based gel-polymer electrolyte to power micro- and meso-scale devices. The inks for each of the cell components − cathode, anode, and electrolytes − are prepared separately. The cathode is composed of 67% MnO_2 and carbon-based ink (6% acetylene black, 8% PVDF-HFP). The anode is composed of either zinc foil, for half-printed cells, or Zn (84%) and carbon ink (7% acetylene black, 9% PVDF-HFP) similar to the cathode ink, for fully printed cells. The gel-polymer electrolyte is composed of PVDF-HFP permeated with the ionic liquid 1-butyl-3-methylimidazolium trifluoromethanesulfonate ([BMIM][Otf]) and the salt zinc trifluoromethanesulfonate (ZnOtf) (67% ZnOtf:[BMIM][Otf], 1:6.5 mass ratio). The fully printed cells and half cells were manufactured with the use of stencil printing (anode and cathode) and doctor-blade coating (GPE) [21]. On cycling, these layer-by-layer printed batteries exhibited significant improvements of discharge capacity, cycle life, and internal resistance over cells previously mechanically assembled by the researchers. These improvements are in large part due to the increased interfacial cohesion between layers, as shown in the scanning-electron-microscopy (SEM) images (Figure 4.1), afforded by direct layer-by-layer printing [21]. The fully printed cells exhibited average discharge capacities of 0.548 mAhcm^{-2}, volumetric energy density of 8.20 mWhcm^{-3}, and gravimetric energy density of 2.46 mWhg^{-2}, with some cells achieving over 1000 cycles without failure.

4.2.2 Spray-printed Electrolytes

Spray printing is used to deposit inks with a wide range of viscosities over various surfaces. The method is based on an airbrush device, in which a high-velocity compressed carrier gas atomizes ink from a reservoir. The brush is directed towards the target substrate. There are a number of ways to control the thickness of a sprayed layer, which include opening the spray nozzle, and varying the pressure of the carrier gas, ink composition, and the number of spray passes. Spray deposition can be used to print a complete battery or certain layers of the battery. Spray printing is attractive for sequentially printing multiple inks that share a common solvent one on top of the other. Because of

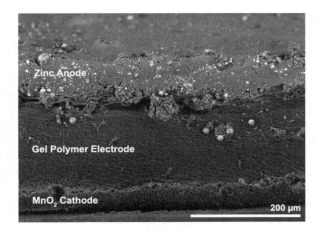

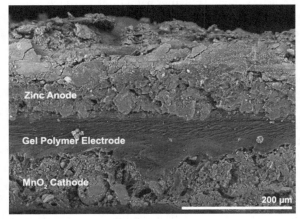

Figure 4.1 Cross-sectional SEM images of fully printed cells with thick and thin GPEs. Left: fully printed thick GPE cell; right: fully printed thin GPE cell [21].

the quick drying of spray-deposited inks, multilayer films can be printed without the necessity to wait for a layer to dry before printing the next layer [20].

Fabrication of battery electrodes and electrolytes by spray printing requires formulation of component materials into liquid dispersions (paints) which can be sequentially coated onto substrates to achieve the multilayer battery configuration. The potential of the fully spray-printed battery remained unfeasible until recently [22–24]. In 2012, Singh *et al.* [22] reported fabrication of full LIBs consisting of $LiCoO_2$ and $Li_4Ti_5O_{12}$ by a spray-printing process. The most important technological achievement of this work was the unique ability to coat the gel-polymer electrolyte, as this had been the bottleneck of the print- or paint-based battery-fabrication process [1]. The polymer separator paint was prepared by dispersing a 27:9:4 (by weight) mixture of KynarRex®-2801, PMMA and fumed SiO_2 in a binary solvent mixture of acetone and DMF (N,N-Dimethylformamide). KynarRex was used because of its good solubility in low-boiling-point solvents and electrochemical stability over a wide voltage window [1], while PMMA was used to promote adhesion to a variety of substrates. Kynarflex-PMMA separators fabricated from paints in acetone had good adhesion, but had a fibrous morphology with very high porosity and excessive electrolyte uptake, which made them mechanically unstable. The authors found that by adding DMF to the paint, the micro-porosity and electrolyte uptake could be tailored to make the separators mechanically robust upon electrolyte addition. This, however, reduced the ionic conductivity of the MGE (microporous gel electrolyte) by a factor of four at 11% DMF content. A further addition of 10% (w/w) fumed SiO_2 to the ink helped offset this loss in conductivity and gave the best compromise between mechanical stability, porosity and ionic conductivity [22]. The membrane was deposited on the previously dried cathode by spraying polymer paint on the surface, which had been preheated to 105 °C, the glass-transition temperature, Tg, of PMMA. During spraying it was necessary to deposit the first few coats slowly and allow them to dry, in order to create an interfacial adhesion layer. Without this step, the separator peeled off the substrate. Subsequent coats of polymer paint were then applied up to a final thickness of 200 μm. Full LIB cells were fabricated by sequentially spraying the component paints on the desired surface with the use of an airbrush. The paints can be sprayed through a set of masks made according to the desired geometry of the device. A cross-sectional SEM micrograph of a spray-printed Li-ion cell (Figure 4.2) [22] shows component layers of reasonably uniform thickness and well-formed interfaces.

Connected in parallel, nine full lithium-ion cells could store total energy of about 0.65 Wh, equivalent to a 6 Whm^{-2} footprint [22, 25] at a thickness of approximately 0.5 mm [1]. The capacities of eight out of nine cells fall within 90% of the targeted capacity of 30 mAh, suggesting good process control over a complex device, even with manual spray painting. Further increasing the energy density could be achieved by making thicker active layers. The current cell retained approximately 90% of its capacity after 60 cycles with a coulombic efficiency of above 98% [20, 22]. This performance is still lower than that of conventional batteries with the same electrodes (Figure 4.3).

4.2.3 Direct-write Printed Electrolytes

Direct writing (DW), also known as digital writing or digital printing, denotes a group of processes which are used to precisely deposit functional and/or structural materials onto a substrate in digitally defined locations. With DW, it is possible to produce precise features from the nm- to the mm-size, because of its conformal-writing capability. The main types of inkjet printing processes, which are related to direct-printing methods, are continuous and drop-on-demand (DOD). As with other printing methods, the final printed

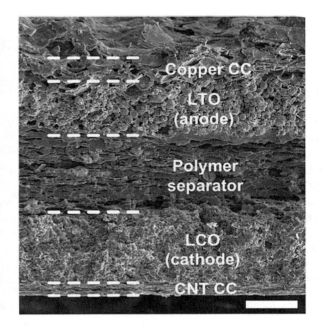

Figure 4.2 Cross-sectional SEM micrograph of a spray-painted full cell showing its multilayer structure, with interfaces between successive layers indicated by dashed lines for clarity (scale bar is 100 mm) [22].

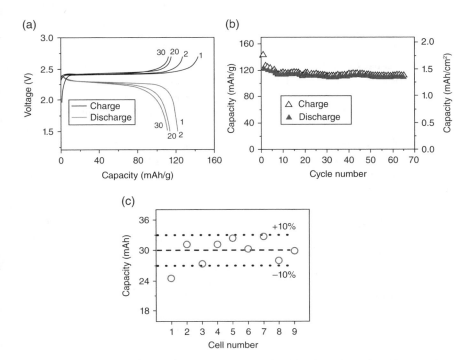

Figure 4.3 Charge–discharge curves for 1st, 2nd, 20th and 30th cycles (a) Specific capacity vs cycle numbers for the spray-painted full cell (LCO/MGE/LTO) cycled at a rate of C/8 between 2.72 and 1.5 V, (b) capacity of the different printed cells, showing 10% variation of the absolute capacity values (c) [20, 22].

shape strongly depends on the ink viscosity, which is a function of the molar mass of the polymer. The printing height depends on the final dried-drop diameter, which is a function of the polymer concentration [26].

The choice of materials is exceptionally wide, ranging from metals and ceramics to polymers and biomaterials. The application of direct-writing techniques requires the development of ink-like mixtures of the battery components with proper colloidal properties in order to make them compatible with the printing machines [27–29]. Slurries used for the inks are typically composed of a polymer, active electrode material and, if necessary, additives. The fraction of the solvent in the ink is kept to a minimum in order to prevent sedimentation, and the active-material particles are ball-milled to reduce the average particle size [20]. To achieve homogeneous inks, rigorous mixing sequences of mechanical shaking and ultrasonication are implemented [30].

Rechargeable zinc/metal oxide (MnO_2) battery utilizing an ionic-liquid-gel electrolyte, prepared by dispenser printing, is presented in [30–32]. Dispenser printing can be used to print inks over areas ranging from $100 mm^2$ to $1 m^2$ by drawing patterns in the form of repeated lines or drops. Because of the non-contact nature of dispenser printing, the ink can be printed over uneven surfaces, something that is not possible with other roll-to-roll printing methods. Reports on batteries fabricated with the use of dispenser printing have predominantly involved printing the active layers and polymer electrolyte onto glass substrates with pre-patterned current collectors, formed by lithography [20].

PVDF-HFP copolymer material is commonly used in batteries and supercapacitors because of its ability to remain structurally robust while absorbing large volumes of electrolytes [30]. The gel electrolyte [30–32] is composed of a 1:1 mixture of PVDF-HFP and 0.5 M solution of zinc trifluoromethanesulfonate ($Zn (CF_3SO_3)_2$, ZnTf) salt dissolved in 1-butyl-3-methylimidazolium trifluoromethanesulfonate (BMIM + Tf-) ionic liquid. This composition was found to have optimal mechanical integrity and ion-transport properties. The viscosity of gel electrolytes is typically an order of magnitude greater than that of neat ionic liquids and both show Newtonian behavior with constant viscosities within the range of the shear rates applied ($0-1 \times 10^4 s^{-1}$) (Figure 4.4).

Therefore, the ionic conductivity of the room-temperature gel ($0.37 mS \cdot cm^{-1}$) is an order of magnitude lower than that of the neat ionic-liquid-based electrolyte ($2.4 mS \cdot cm^{-1}$). However, the gel is considered fairly conductive compared to dry solid-polymer PEO-based electrolyte [33] ($0.01 mS \cdot cm^{-1}$) and glassy LiPON at room temperature electrolytes ($<10 \mu S.cm^{-1}$) [34]. To fabricate thin samples, the PVDF-HFP-based gels were printed on glass substrates with 30-gauge ($150 \mu m$ inner diameter) needles, and dried on a hotplate at $60 °C$ for 45 minutes. In [31] it was found that the resulting microstructure of the gels is scarcely affected by printing parameters such as needle dimension, film dimensions, print speed and substrate material, and more dependent on the gel composition and post-process conditions, such as the drying temperature. The test

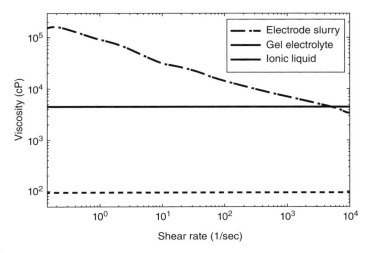

Figure 4.4 Viscosity behavior of the representative electrode, gel electrolyte and ionic liquid inks as a function of the applied shear rate [30–32].

cells were printed with $0.25\,cm^2$ footprint areas and total thicknesses of between 80 and 120 μm. Multiple films were successively deposited to form a stacked microbattery configuration. Electrolyte thickness varied from 15 to 30 μm [32]. As can be seen from the linear-sweep voltamogram (Figure 4.5), the electrolytes are stable between 0 and 2.7 V with respect to zinc. This potential range is sufficient for the battery operating between 1 V and 2 V [32].

The compatibility of the gel electrolyte with the zinc electrode is critical to the battery's performance. It was found that the interfacial resistance between the electrode and the gel electrolyte increases rapidly within the first hours after assembly, and is stabilized in 24 hours (Figure 4.6). Cyclic-voltammetry measurements show that the passivation layer does not block the reversible dissolution and deposition of zinc. Over many cycles, these reactions occur reversibly with similar current densities. Therefore, it can be concluded that zinc showed good compatibility with the gel electrolyte, and when cycled against a manganese-dioxide electrode. The $0.25\,cm^2$-footprint cell exhibiting storage capacities of $0.98\,mAh/cm^2$ ran over more than 70 cycles. Energy density of $1.2\,m\ Whcm^{-2}$ was measured [30].

It is rather well established that the corrosive nature of traditional acidic and alkaline aqueous electrolytes, and challenges with shape changes at the zinc electrode–electrolyte interface have prevented commercial zinc-based batteries from being used as rechargeable [35]. The development of the gel ionic-liquid electrolyte has eliminated this disadvantage.

Thin and flexible Zn/MnO_2 batteries printed and laminated on thin paper have already been commercialized by several companies (Blue Spark, Power

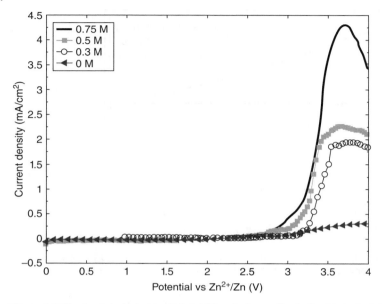

Figure 4.5 Electrochemical potential stability windows of ionic-liquid electrolytes with 0–0.75 M zinc salt concentrations [32].

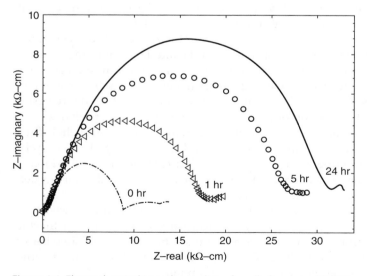

Figure 4.6 Electrochemical impedance plots of a cell after assembly [32].

Paper, Enfucell, etc.) and are used for disposable applications including RFID, smart cards, microsensors, functional packaging and media, etc. [36]. Blue Spark, Power Paper and Enfucell batteries are carbon/zinc primary printed batteries with alkaline or neutral (zinc chloride) aqueous electrolyte.

Recent progress in the high ionic conductivity (10^{-3} S.cm^{-1}) of zinc polymer electrolytes has enabled the production of fully printable cell structures that do not use a separator. The polymer electrolyte layer solidifies from solution during drying into a self-supporting film after being deposited directly onto the underlying electrode. This approach simplifies the manufacturing process and also allows for a mechanically integrated cell stack, as each adjacent layer is bonded in a monolithic stack of successive solution-printed layers. The method enables the creation of a cell stack that resists delamination and is robust under flexing, as has been shown by dynamic flexibility testing of these printed cells [35].

ZincPoly battery technology leverages cost-effective, stable and earth-abundant zinc and manganese-dioxide battery electrodes that have been engineered to be printable and compatible with the novel high-ion-conductivity solid-polymer electrolyte. Two-hundred-cycle rechargeability with less than 10% capacity loss was measured in the cell with printed zinc polymer electrolyte at a C/2.5 discharge rate. This high cycling stability for a secondary zinc battery is thought to be due to stable plating of the zinc ions from the electrolyte during recharging, since metallic dendrites or mossy deposits are suppressed. Three-dimensional anode-interface growth can cause electrode short circuits and degrade cell-cycling stability. A scalable, large-area print process has been developed for zinc polymer batteries. The process has been optimized with regard to print parameters, compositions and film thicknesses to reach target volumetric and areal energy densities and capacities [35]. Figure 4.7 from [35] shows continuous high-discharge-rate data and ragone plot from $10\,cm^2$ zinc polymer cells which can be used to power a BTLE (Blue Tooth Low Energy) wearable wireless sensor. These cells are fabricated with real impedances in the range of 2–5ohms at 1kHz. In terms of capacity, less than 500 micron-thick batteries based on this technology demonstrated capacities of above 5mAhcm^{-2} when discharged from 1.8 to 1 V.

Table 4.1 compares the capabilities of printed-zinc-polymer technology to lithium and conventional zinc primary technologies. Thin secondary zinc batteries could be produced at approximately 50–60% of the cost of lithium-polymer cells in the 50–100mWhcm^{-3} capacity range. The printed-zinc-polymer technology has been scaled to a production capacity of one thousand cell prototypes and is scheduled to be scaled to pilot production in early 2017.

Wright *et al.* [18] demonstrate an on-glass dispenser-printed lithium-ion battery with solid-polymer electrolyte. The electrolyte contains LiCF$_3$SO$_3$N (LiTFSI) salt with butyl-methylimidazolium [BMIM]$^+$ and butyl-methyl pyrrolidinium [BMPyrro]$^+$ ionic liquid. These ionic liquids have been shown to form solid films or self-standing gels in a mixture with PVDF, maintaining good ionic conductivity. The electrolytes were printed with the use of the

(a)

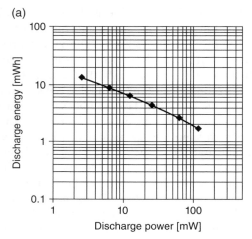

(b)

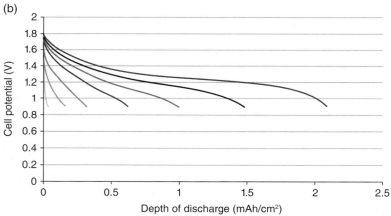

Figure 4.7 (a) High-discharge-rate data from printed 350-μm-thick zinc polymer battery with more than 5C-rate discharge capability. (b) Discharge curves at continuous currents (0.2, 0.5, 1, 2, 5, 10, 20 mA/cm² (blue through gray) from printed zinc polymer cell (the data in (a) and (b) are obtained from different cells) [35].

homemade pneumatic printer [18]. The details are as follows: Pyrrolydinium-TFSI salts were mixed in 1:1 (w/w) PVDF powder. LiTFSI salt was added after formation of a viscous gel. The IL/PVDF/LiTFSI layer was applied on top of the anode with a 100 μm stainless-steel needle. The gel was baked under a heat lamp for 10 minutes. By carefully controlling the dispenser pressure (40 to 70 kPa) the 12 μm–thick layer was obtained. The printed polymer electrolyte exhibited electronic insulation, high ionic conductivity ($2 \times 10^{-3} \text{Scm}^{-1}$ at 30 °C (Figure 4.8)) and mechanical stability [18].

Table 4.1 Summary of candidate flexible battery technologies.

	Lithium ceramic composite and thin film lithium	Thin zinc carbon primary	Lithium polymer pouch cell	Printed zinc polymer
Format	Flex PCB laminate and semi-rigid thin PCB and coated films	Assembled separator + printed components in sealed pouch laminate	Assembled electrolyte separators and electrodes in vacuum-sealed pouch cell	Fully printed cell stack with laminated packaging
Energy Density in Thin Format	<50–150 Wh/L	<50–150 Wh/L	100–200 Wh/L	100–300 Wh/L
Cycle Life	>500 cycles	1	>500 cycles	200 cycles
Pulse Capability	3–10C	C/3 to C/20	1–5C	5–10C
Flexibility	Poor for vacuum deposited cells. OK for ceramic composite cells	Poor to fair. Package buckling an issue in some designs	Poor. Packaging failure, gas expansion and electrical failure	Good. Monolithic, flexible polymer-based structure
Safety	Fair. Reduced leakage and expansion versus lithium polymer cells	Good. Low-toxicity and low-reactivity anode and cathode materials	Poor. Flammability, reactivity, disposal concerns	Good. Low-toxicity, low-volatility electrolyte and low-reactivity anode and cathode materials

An inkjet-printed zinc/silver battery has been reported with an energy density of $3.95\,\text{mWh.cm}^{-2}$ [16]. A battery was constructed by submerging a pair of silver inkjet-printed structures into an aqueous electrolyte of potassium hydroxide (KOH) with dissolved zinc oxide (ZnO) powder. The optimal molarity of KOH and the concentration of dissolved ZnO were investigated. In order to replace liquid alkaline electrolyte, an alkaline gel electrolyte has been developed. The electrolyte composition is optimized with respect to polymer material (polyethylene oxide (PEO and methylcellulose)), polymer molecular weight, and $KOH:H_2O:PEO$/cellulose ratio [37]. On the one hand, PEO is known as a poor ion-conducting polymer at room temperature. The reported conductivities are in the order of $10^{-7}\,\text{S.cm}^{-1}$ [37]. On the other hand, PEO ionically crosslinks with sodium ion in alkaline electrolyte and reduces the

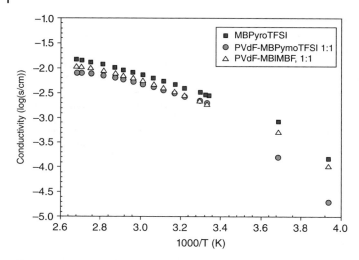

Figure 4.8 Ionic conductivity of PVDF/ionic-liquid mixtures [18].

corrosion of zinc. Methylcellulose was tested as one of the gel-electrolyte components on the basis of its use as a separator in alkaline battery systems. Alkaline gel electrolytes were fabricated by a two-step method. First, a polymer solution was printed by extrusion on top of the electrodes. This was followed by pipetting the appropriate amount of potassium hydroxide solution [37]. A planar printed primary 2D silver/zinc battery with gel electrolyte was found to deliver areal energy density of 4.1 ± 0.3mWh.cm^{-2} [37].

Clarka *et al.* developed a flexible zinc-air battery that was produced by screen-printing an anode based on zinc/carbon/polymer composite, and a vapor-polymerized PEDOT cathode on two sides of photo-quality paper [38] and applying a lithium chloride electrolyte between the two electrodes. For printing the electrolyte, an aliquot of the LiCl/LiOH solution was mixed with molten ($\sim$60 °C) polyethylene glycol (PEG, MW 1500), in a 1:1 ratio by weight, to make 6 M LiCl:P(EO). Ethanol or butan-1-ol was added to obtain 10 mPa.s viscosity for inkjet printing [38]. To ensure sufficient coverage, eight layers of the electrolyte were printed on the substrate at a drop spacing of 20 μm and a cartridge temperature of 30 °C. Evaporation of the solvents caused the electrolyte to solidify. In other experiments, 8 M LiCl/LiOH electrolyte was formulated without the addition of PEG, and ethanol or butan-1-ol was added to lower the solution viscosity for inkjet printing [38].

As already mentioned, printing of the electrolyte layer for the rechargeable printed battery is even more challenging than for the primary battery, since the electrolyte/membrane must withstand continuous volumetric changes of electrode materials occurring on charge/discharge and enable longevity of the battery. A printable paste formulation for the electrolyte has been developed in [39]. In order to bring the 25% caustic-potash-solution electrolyte into screen-printable form, different binder/solvent combinations and thickener agents were tested. This electrolyte paste then also acts as separator [39]. The specifications for the Zn/MnO_2 cell are as follows: $20 \times 20 \, mm^2$ electrode, 1 mA charging/discharging current and 20mAh capacity. The capacity of the NiMH-cell of the same size varies from 16 to 32mAh.

The major hurdles in formulating printable gel electrolytes that remain stable over the battery life are addressed by Gethin *et al.* [39]. The authors suggest a roll-to-roll (R2R) process for overcoming some of the challenges of fabrication of flexible printable wearable batteries. The electrodes were screen-printed while the gel electrolyte was sandwiched between them. The screen-printing process enabled the sequential deposition of current collector, electrode and separator/electrolyte materials on a polyethylene terephthalate (PET) substrate in order to form flexible and rechargeable electrodes for battery application. The anode and cathode were based on the conducting poly(3,4-ethylenedioxy-thiophen), poly (styrene sulfonate) (PEDOT, PSS) and polyethyleneimine (PEI), which were combined to form the electrodes. A layer of electrolyte (sodium styrene sulfonate – PSSNa) in the form of a gel or a precast film with a thickness of about 3 μm was eventually placed and not printed over the electrode, covering its entire surface [39]. The difference in the redox potential between the two electrodes produced an open-circuit voltage of 0.60 V and the battery displayed a specific capacity of $5.5mAh.g^{-1}$, which is comparable to the specific capacities obtained from conventional rechargeable lead/acid batteries. The battery developed had an active surface area of $400 \, mm^2$ and a device thickness of 440 microns. As this is a thin-polymer battery and the amount of active material is very small, this battery can serve only limited applications, such as powering devices that are activated periodically for short times.

The ionogel inkjet printing method is used by Delannoy *et al.* [27, 40] to develop a fast deposition process of solid electrolyte for a microbattery. The authors [40] combine inkjet printing with sol-gel technique to obtain ionogel electrolytes. In the sol-gel process, the precursors for preparation of colloid consist of a metal or metalloid element surrounded by various ligands. The polymerization takes place as a result of a condensation reaction between two hydrolyzed precursor molecules. The most common examples of precursor are tetraethoxysilane (TEOS) and tetramethyl orthosilicate (TMOS). The sol precursor of the ionogel was prepared by mixing tetramethyl orthosilicate (TMOS), the ionic liquid N-methyl-N-propylpyrrolidinium bis(trifluoromethane)sulfonylimide (PYR13-TFSI) and the lithium salt lithium bis(trifluoromethane)sulfonylimide (LiTFSI). The IL/TMOS, and IL/TEOS, molar ratios are 0.25 and 1 [40]. The viscosity of the ionogel precursor (approximately 10mPa.s) is appropriate for printing. The deposition was directly performed onto the porous composite electrodes ($LiFePO_4$ and $Li_4Ti_5O_{12}$ cathode- and anode-active materials respectively). Droplets from the cartridge tank are ejected on demand through the nozzles to form the desired pattern. The ionogel conductivity is found to be two orders of magnitude higher than that of LiPON for an IL/TMOS molar ratio of 0.25.

A study of the conductivity of ionogels based on confined-in-silica ionic liquids has been published in [41]. For 1:1 IL:TMSO the ionogel shows conductivity similar to that of PYR13-LiTFSI liquid electrolyte. This is only one order of magnitude lower at room temperature than that of the standard nonaqueous liquid electrolyte of Li-ion batteries (1 M LiPF$_6$ in the EC:DMC binary solvents (LP30)). Half cells and complete batteries have been assembled with tape-cast LiFePO$_4$ positive and Li$_4$Ti$_5$O$_{12}$ negative porous composite electrodes, both coated by an ionogel solid-electrolyte layer. The coated electrodes are inserted, facing each other, into a Swagelok cell. A fiberglass separator, soaked in PYR13-TFSI and LiTFSI solution, is inserted in a way that makes good wetting contact between the two solid-electrolyte ionogel surfaces. The first discharge of complete cells shows a capacity of 125 mAh.g^{-1} at C/10. The discharge capacity remains above 100mAh.g^{-1} after 25 cycles (Figure 4.9). Areal capacity of 0.3mAh.cm^{-2} is obtained, which corresponds to a specific capacity of 60mAh.g^{-1} for the LiFePO$_4$ positive electrode. By comparison, capacities obtained with LP30 liquid electrolyte are 145mAh.g^{-1} and 0.75mAh.cm^{-2} at a C/30 rate. Such differences are related to different ionic conductivities,

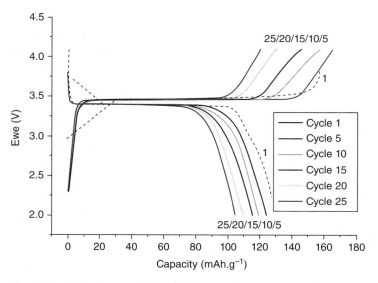

Figure 4.9 Galvanostatic cycling of tape-cast LiFePO$_4$-based porous composite electrode with printed ionogel in half-cell with lithium-metal counter electrode from 2.0 to 4.1 V at C/30 rate and at room temperature [41].

viscosities and interfacial problems. On the other hand, a capacity close to 0.3mAh.cm^{-2} is much higher than that of microbatteries obtained by PVD (0.12–$25 \,\mu\text{Ahcm}^{-2}$) [40, 42, 43]. Cycleability appears to be good over 100 cycles.

On the basis of the data presented above, particularly on the thickness of electrodes, printed batteries can be considered as power sources intermediate between conventional bulk batteries and thin-film batteries. The 3D-printed battery (3D-PB) is one of a series of printed batteries. Its structure is more complex than that of the 2D-printed battery, since 3D-PB implies high-aspect-ratio, complex-shape electrode architectures. As with 2D-battery printing, in the 3D printing process of a battery, the most challenging part is the printing of the solid electrolyte, a problem as yet unresolved.

Hu *et al.* [24] have recently developed graphene oxide (GO)-based electrode composite and solid membrane inks to print by DIW (direct ink writing) the all-component 3D lithium-ion battery with interdigitated electrodes. The inks developed are aqueous GO-based electrode slurries, consisting of high-concentrated GO with cathode- or anode-active materials. The GO contributes to the formation of suitable rheological properties of the ink, which enable 3D printing. The electrode filaments are extruded directly from a nozzle and deposited layer by layer. The dimensions of the interdigitated electrode are $7 \,\text{mm} \times 3 \,\text{mm}$ with a six-layer thickness. As a result of the shear stress induced by the nozzle, the GO flakes are aligned along the extruding direction, which enhances the electrical conductivity of the electrode [44]. The membrane-ink composite consisting of PVDF- *co* -HFP and Al_2O_3 nanoparticles is printed into the channels between the electrodes. When the sample is dried, the liquid electrolyte is injected into the channel to fully soak the electrodes. An entirely 3D-printed $LiFePO_4/Li_4Ti_5O_{12}$ full cell features a high electrode mass loading of about $18 \,\text{mg.cm}^{-2}$ when normalized to the overall area of the battery. The full cell delivers initial charge and discharge capacities of 117 and 91mAhg^{-1} with good cycling stability (Figure 4.10).

The current research in our laboratory is focused on the development of an extruder that will create custom-made filaments composed of active anode, cathode and electrolyte materials to simultaneously print all the components of a high-aspect-ratio 3D-microbattery.

4.2.4 Laser-printed Electrolytes

Several laser-printing techniques have been tested for the printing of electrolytes. Matrix-assisted, pulsed-laser evaporation (MAPLE) has been used to deposit a wide variety of materials, including metallic, ceramic, polymeric and even biological materials. However, as with hot-melt inkjet printing, MAPLE is limited to patterning two-dimensional layers or simple three-dimensional structures, since the impinging molten material requires an underlying substrate for support.

(a) (b) (c)

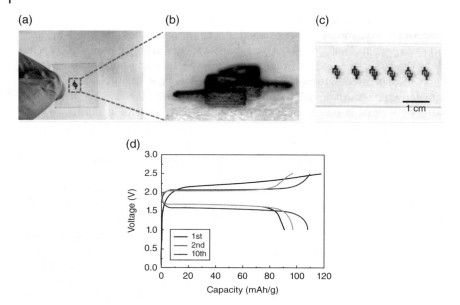

(d)

Figure 4.10 3D-printed miniature cell. Digital images of the 3D-printed electrodes (a) and (b), and arrays (c). Charge and discharge profiles of the 3D-printed full cell (d) [44].

The direct laser-writing technique by multi-photon polymerization, because of its unique properties and characteristics, has proven to be an indispensable tool for high-accuracy structuring and has been put on the map as an emerging technology for scaffold 3D printing. Direct laser writing (DLW) by multi-photon polymerization (MPP) is a three-dimensional printing technology that allows the construction of readily assembled structures with sub-100 nm resolution. It is based on the nonlinear photon absorption by photopolymers; the beam of an ultra-fast laser is tightly focused inside the volume of a transparent material, causing it to absorb two or more photons and polymerize locally. DLW, along with classic stereolithography and selective laser sintering, make up a versatile class of laser-based 3D printing techniques. In general, a material suitable for structuring with DLW includes at least two components: (i) a monomer, or a mixture of monomers/oligomers, which will provide the final polymer and (ii) a photoinitiator, which will absorb the laser light and provide the active species that will cause the polymerization. These are mainly negative photoresists, such as hydrogels, acrylate materials, the epoxy-based photoresist SU-8 and hybrid materials [45].

Selective laser sintering (SLS) creates three-dimensionally patterned materials by locally fusing polymer powder through laser writing. This approach was invented by Deckard [46] and developed by Bourell *et al.* [47]. It shares a common feature with 3DP in that a layer of powder is spread on a piston and selected regions are bonded together. But where 3DP relies on binder deposition, SLS induces the desired powder adhesion by melting (or viscous sintering). Metals, ceramics, glasses, and polymers can be incorporated into the powder system, but a thermoplastic polymer is needed to bind the materials together. The powder size and melting process limit SLS to structures with feature sizes of ~100 μm or more [48]. Stereolithography (SLA) technology employs liquid ultraviolet-curable photopolymer resin and an ultraviolet laser to build the object's layers one at a time. For each layer, the laser beam traces a cross-section of the part pattern on the surface of the liquid resin. Exposure to the ultraviolet laser light cures and solidifies the pattern traced on the resin and joins it to the layer below. This is the greatest difficulty in designing concentrated inks suitable for direct writing on the microscale. Nature has provided the only example of "direct ink writing" of complex structures with micrometer-sized, self-supported features, in the form of spider webs. This low-viscosity fluid undergoes solidification as it flows through a fine-scale spinneret to form silk filaments. One of the advantages of the laser-induced forward-transfer (LIFT) process is that because of its nozzle-free nature, LIFT is capable of printing patterns of 3-D pixels (or voxels = VOlume piXEL) with nano-inks of much higher viscosity than that of inkjet printing, thus significantly reducing variations between patterns due to wetting and drying effects [49]. In the multicomponent liquid or gel systems, the desired material is suspended or dissolved in a liquid to form an ink, which is then spread on the substrate [50]. As with the inkjet approach, thick layers can be deposited and stacked. The use of inks enables the transferred materials to flow and coalesce, forming uniform and continuous coatings upon reaching the acceptor substrate. By optimizing laser and ink parameters, feature sizes on the order of 1–2 μm can be achieved.

One of the important attributes of DLW printing in the context of electrochemical systems is that it allows for the deposition of highly porous, multicomponent materials without modifying their properties. In all cases, the technique results in uniform transfer of the structurally complex materials with a porous structure that allows for good electrolyte penetration. Another key advantage of DLW printing in constructing electrochemical cells is the flexibility in the design of operating geometries.

The viscosity of the ink affects the film thickness, i.e., for high-viscosity inks, thicker films are deposited per DLW pass than for lower-viscosity inks. As the viscosity of the ink increases, the laser power required for its printing increases as well. This increase in laser power not only causes deterioration of the printing resolution, but can also alter the materials suspended in the ink. Thus, optimization of the overall transfer process must also take into account how the higher laser power affects the chemical and material properties of the ink components [49].

Alberto Piqué *et al.* [51–53] use a patented laser direct-lift write printing (LIFT) technique that was developed and optimized at the Naval Research

Laboratory for the deposition of solid-state ionic-liquid electrolyte membranes. The solution was composed of a 1 M lithium salt (lithium-bis-trifluoromethanesulfonylimide) dissolved in an ionic liquid (1,2-dimethyl-3-n-butylimidazolium-bis-trifluoromethanesulfonylimide—DMBI + TFSI-), a polymer (poly(vinylidene fluoride-co-hexafluoropropylene)—PVDF-HFP), a solvent (dibasic ester—DBE), and ceramic nanoparticles (25 nm TiO_2). The composition of the solution was 6.4 wt.% DMBITFSI, 3.6 wt.% PVDF-HFP, 1.1 wt.% TiO_2, and 88.9 wt.% DBE. The suspension is spread on a borosilicate glass slide with a wire coater. Then an $Nd:YVO_4$ laser ($\lambda = 355$ nm) with a μm^2 spot size is used to irradiate the back side of the borosilicate glass slide and transfer the ink onto the substrate below ((~100 μm gap). Once transferred, the resulting ink layer is heated in a convection oven at 75 °C for 60 minutes and then placed in a vacuum oven and heated at 80 °C for 24 hours in order to form a continuous, pinhole-free, moisture-free ionically conductive membrane. Other important properties for solid-state-electrolyte membranes to possess are mechanical strength and flexibility. Figure 4.11 (a, b, c) [52] shows three optical micrographs of a laser-printed solid-state-electrolyte membrane deposited on a glass substrate. In Figure 4.11a, the optical micrograph shows the undisturbed laser-printed membrane. Figures 4.11b and 4.11c show the solid-state-electrolyte membrane being partially and then completely, respectively, lifted from the glass substrate with the use of a pair of tweezers. The membrane thickness is approximately 5 μm. These optical micrographs show that deposited material

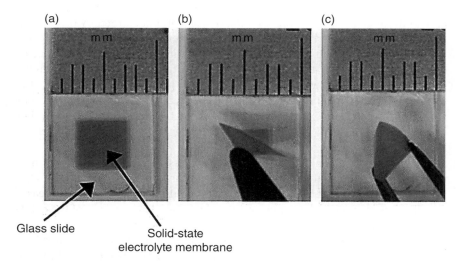

Figure 4.11 Optical micrographs showing the strength and flexibility of the nanocomposite solid-state electrolyte membranes laser printed on a glass slide. (a) Membrane laser printed on a glass slide, (b) membrane partially lifted off from glass slide, and (c) membrane held using tweezers [52].

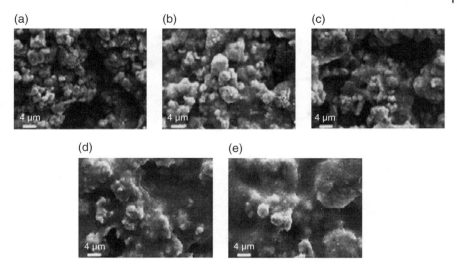

Figure 4.12 SEM micrographs of laser-printed solid-state electrolyte onto porous $LiCoO_2$ cathodes. The inks of various viscosities (DBE concentrations) are: (a) 91 wt.% DBE, (b) 90 wt.% DBE, (c) 89 wt.% DBE, (d) 88 wt.% DBE, and (e) 87 wt.% DBE [52].

forms a continuous, flexible, pinhole-free membrane and that even at a thickness of 5 μm, the membranes are strong enough and flexible enough to be peeled off with tweezers without causing the membranes to tear or form cracks.

The effect of the viscosity of the electrolyte suspension on the resulting solid IL membrane was tested. Figure 4.12 [52] shows scanning-electron-microscopy (SEM) micrographs of different-viscosity solid electrolytes deposited on a cathode material ($LiCoO_2$) commonly used for Li-ion battery systems. As the percentage of dibasic ether solvent (DBE) decreases (from 91 wt.% in (a) to 87 wt.% in (e)), one can see that the separator does not seep as far down into the porous cathode ($LiCoO_2$). By controlling the viscosity of the precursor electrolyte suspension, one can begin with a low-viscosity suspension that allows the electrolyte material to percolate and seep down into the pores of the cathode and anode to gain access to all of the active material. This is then followed by a high-viscosity suspension that will form a solid separator layer on the surface of the cathode that isolates the cathode from the anode. In addition, the higher viscosities help to planarize the porous cathode surface in order to reduce the chances of short circuits when the anode is deposited.

The nanocomposite solid-state electrolyte films exhibited ionic conductivity of about $1.5 \, ms \cdot cm^{-1}$ at about 30 °C, (Figure 4.13) [52]. Higher values in the range of 1.8 to 3.1 $mS \cdot cm^{-1}$ were reported elsewhere [54] as a function of TiO_2 and of the amount of ionic liquid. These ionic conductivities are typical of ionic-liquid-based electrolytes, namely, 2–5 $mS \cdot cm^{-1}$ [51]. These results show that the process of laser printing does not damage the ionic conductivity of the transferred material.

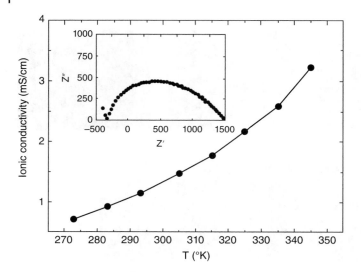

Figure 4.13 The temperature dependence of ionic conductivity of the laser-printed nanocomposite solid-state electrolyte membrane. The inset figure is the complex resistivity plot for the same membrane [52].

A lithium microbattery was assembled using the laser-printed polymer membrane, which was sandwiched between a laser-printed $LiCoO_2$ cathode and a lithium-metal anode. This microbattery was cycled by chronopotentiometry at a C/18 rate. The footprint of the active area was $9\,mm^2$ with an active mass (cathode-limited) of 483 mg. The separator thickness was approximately 5 µm. The microbattery had a charge/discharge efficiency of ~98% and a capacity per area of 205mAhcm^{-2}. Li-ion microbatteries were also fabricated with these ~20 µm solid-polymer membranes with the use of an $LiCoO_2$ cathode (30 µm) and a carbon anode [53]. The battery was charged and discharged at a constant current density of 40 µA/cm^2 (0.1C rate) between 4.2 V and 3 V. After the fourth cycle, the microbattery had a coulombic efficiency of 98% with a discharge capacity of 0.495mAhcm^{-2}. Both capacity values (of lithium and lithium-ion cells) are higher than the value of 0.16mAhcm^{-2} achieved for sputter-deposited lithium microbatteries reported by Bates *et al.* [52].

Another work by Alberto Piqué *et al.* [54] demonstrates the deposition of sequential layers by LDW (laser direct writing) of an $LiCoO_2$ cathode, a composite solid-polymer ionic liquid (c-SPIL) electrolyte and a carbon anode. The composition of the c-SPIL electrolyte is similar to that employed in previous work, except that the amount of DBE is not specified [52]. All layers are deposited in a $3 \times 3\,mm^2$ laser micro-machined pocket on a polyimide/aluminum

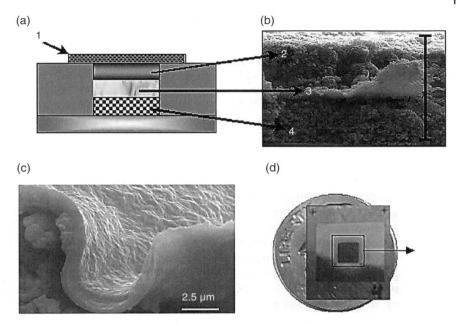

Figure 4.14 (a) Representative cross-section and cutout schematic of an embedded lithium-ion microbattery fabricated by laser direct-write (LDW) printing using a nanocomposite solid-polymer ionic-liquid (nc-SPIL) electrolyte. (b) Scanning-electron-microscopy (SEM) cross-section showing the layer structures. The numbers 1 through 4 correspond to the metal current collector, the carbon anode, the nc-SPIL electrolyte, and the lithium cobalt oxide cathode, respectively. Layers 2, 3, and 4 are deposited sequentially by LDW. (c) SEM images of the nc-SPIL separator with the cathode and anode removed, showing the structural integrity of the nc-SPIL after cleaving. (d) Actual LDW microbattery in a polyimide substrate shown against a US dime for scale. The black square (3 mm^2) indicates the active battery portion of the system [54].

flexible substrate. The overall thickness of these cells is approximately 30 μm, corresponding to a total volume of 270 nL. The microbatteries are cycled more than 100 times between 4.65 and 3 V at charging and discharging currents of 0.11 mA/cm^{-2}.

SEM images (Figure 4.14) [54] show that the LDW-transfer process deposits the c-SPIL ink to form a uniform, pinhole-free solid-polymer layer. Because of the granular nature of the cathodic layer, the initial laser-deposited c-SPIL separator is able to partially penetrate this material, allowing for enhanced surface-area interactions between the electrolyte and the cathode. The LDW c-SPIL electrolyte composed of 30 wt% PVDF-HFP, 52 wt% 1 M Li/BMMITFSI, and 18 wt% of 60–80 nm to 1–2 μm-size barium titanium oxide particles (BTO) has an ionic conductivity in the range 0.02–1.69 mS·cm^{-1}

Table 4.2 Dependency of room temperature ionic conductivity on the size of ceramic particulates (barium titanium oxide) added [54].

Average BTO particle size	Ionic conductivity (mS/cm) at 22 °C
1–2 μm	0.02588
600 nm	0.53093
60 nm	1.69011

(Table 4.2). The data clearly indicate that the addition of smaller particles results in higher ionic conductivity [51, 54]. This size dependence is similar to the behavior seen in PEO-based composite polymer electrolytes [51] and thus is not related to the method of preparation, i.e., printing or casting. The dependence of the ionic conductivity on the amount of TiO_2 ceramics was tested. The measured room-temperature ionic conductivity of 2.8–3.11 mS·cm^{-1} (10–18 w/w% TiO_2) is close to that of typical ionic-liquid electrolytes. It is more than two orders of magnitude higher than that of LiPON [54] and 1 to 4 orders of magnitude higher than those of solid amorphous or crystalline polymer electrolytes. The TiO_2 nanopowder serves as a stiffener of the nanocomposite and, in addition, partially absorbs the incident laser radiation during the LIFT process.

Figure 4.15 [54] illustrates the charge–discharge (C/3 Rate, 9.9 μA at 0.11mAcm^{-2}) of the Li-ion microbattery with an $LiCoO_2$ cathode and carbon

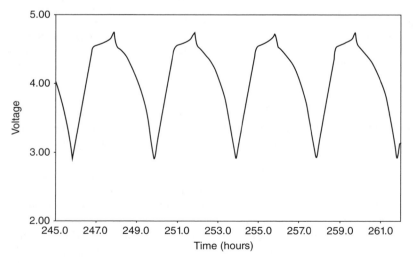

Figure 4.15 50th to 53rd charge–discharge cycle of an $LiCoO_2$-based microbattery made by LDW [54].

anode. The energy density for this system is $0.132 mWh cm^{-2}$ (and $0.406 mWh cm^{-3}$) corresponding to a specific energy of $0.33 mWh g^{-1}$ on the basis of the battery mass. The cathodic capacity of this battery is near $100 mAh g^{-1}$. Thus, these LDW microbatteries exhibit energy densities and capacities similar to those observed for typical lithium-ion batteries and state-of-the-art microbatteries [54].

Nanoimprint lithography (NIL) is a novel method of fabricating micro/nanometer-scale patterns at low cost, with high throughput and high resolution. Unlike traditional optical-lithographic approaches, which create patterns through the use of photons or electrons to modify the chemical and physical properties of the resist, NIL relies on direct mechanical deformation of the resist and can therefore achieve resolutions beyond the limitations set by light diffraction or beam scattering that are encountered in conventional lithographic techniques [55]. Lee *et al.* use the UV-IL technique to construct 3D shape-conformable polymer electrolytes featuring a maze-like structure with a repeating surface grating (wall thickness and height of $10 \mu m$) in $1.5 cm \times 1.5 cm$ footprint. Polydimethylsiloxane (PDMS) was pressed against cast layers of precursors, composed of ethoxylated trimethylolpropane triacrylate (ETPTA) monomer, photoinitiator, and alumina nanoparticles. UV exposure through the stamps yielded solid replicas of composite gel-polymer electrolyte. Cycling of the $LiCoO_2$/composite GPE/Li cell exhibits stable charge/discharge profiles up to the 50th cycle with coulombic efficiency of 97% [56].

4.3 Summary

While the field of printed batteries is newly developed, it has progressed greatly within the last few years. The printing of ion-conducting membranes and electrolytes remains a bottleneck in all-printed devices, because of the diverse, stringent requirements imposed on electrolytes. In order to enable high performance of a printed battery, high ionic conductivity, and compatibility with electrode materials are essential properties. In addition, the printing of electrolyte directly onto the electrode surface is very problematic. Development of electrolyte inks is the major challenge in the fabrication of an all-printed battery. The composition and rheology of each ink must be optimized in order to ensure reliable flow, to promote adhesion between the printed features, to provide the structural integrity needed to withstand drying and sintering without delamination or distortion, and above all, to maintain the electrochemical activity of the electrode and electrolyte materials. A very few publications appear in this field, with polymer-gel or swelled membranes as the most studied. Zn/MnO_2, Zn/Ag, Zn/air, $LiCoO_2/Li_4Ti_5O_{12}$ and $LiFePO_4/Li_4Ti_5O_{12}$ are the major types of all-printed batteries developed up to now. Polymers used for membranes and gel electrolytes in these batteries include PVDF-HFP, PMMA, PEO, PAN, methylcellulose and PDMS. Alumina, silica and titanium oxide nanoparticles are used as ceramic fillers to increase the mechanical strength and uptake of alkaline aqueous or aprotic carbonate, γ-butyrolactone or ionic-liquid-based electrolytes. Direct, inkjet, screen or laser printing and stencil brush or spray-painting methods are applied for the printing of gels and membranes. The ionic conductivity of printed electrolytes was found to be close to the conductivity of the counterpart conventional battery electrolytes. The electrochemical performance of printed batteries approaches and in some cases outperforms that of conventional bulk and thin-film batteries. However, as opposed to the latter, there are no publications which deal with investigation of ion-conduction mechanisms in printed electrolytes and interfacial electrode/electrolyte phenomena. No experiments and results have been published up to now on the printing of all-solid-state polymer and ceramic electrolytes.

We believe that owing to the great promise latent in the concept of the fully printed battery, the intensive research effort that must be devoted to the development of printed-electrolyte technology will be valuable both to electrochemical science and to the battery market. The most progressive 3D-printing technique, which enables simultaneous printing of all battery components, is expected to contribute greatly to electrochemical energy-storage devices.

References

1 Vlad, A., Singh, N., Galande, C., Ajayan, P.M. (2015) Design considerations for unconventional electrochemical energy storage architectures. *Adv. Energy Mater.* **5(19)**.

2 Seki, S., Ohno, Y., Kobayashi, Y., Miyashiro, H., Usami, A., Mita, Y. *et al.* (2007) Imidazolium-based room-temperature ionic liquid for lithium secondary batteries effects of lithium salt concentration. *J. Electrochem. Soc.* **154(3)**, A173–A177.

3 Reddy, T. (2010) *Linden's Handbook of Batteries.* 4th edition, McGraw-Hill.

4 Franco, A.A. (2015) Electrolytes for rechargeable lithium batteries. In M. Montanino, S. Passerini, G.B. Appetecchi (eds) *Rechargeable Lithium Batteries: From Fundamentals to Applications*, Woodhead Publishing Series in Energy: Number 81, Amsterdam.

5 Etacheri, V., Marom, R., Elazari, R., Salitra, G., Aurbach, D. (2011) Challenges in the development of advanced Li-ion batteries: a review. *Energy Environ. Sci.* **4**, 3243–3262.

6 Xu, K. (2004) Nonaqueous liquid electrolytes for lithium-based rechargeable batteries. *Chem. Rev.* **104**, 4303–4418.

7 Idris, N.H., Rahman, M.M., Wang, J.Z., Liu, H.K. (2012) Microporous gel polymer electrolytes for lithium rechargeable battery application. *J. Power Sources* **201**, 294–300.

8 Prasanth, R., Aravindan, V., Srinivasan, M. (2012) Novel polymer electrolyte based on cob-web electrospun multi component polymer blend of polyacrylonitrile/poly (methyl methacrylate)/polystyrene for lithium ion batteries—Preparation and electrochemical characterization. *J. Power Sources* **202**, 299–307.

9 Choi, B.K., Shin, K.H., Kim, Y.W. (1998) Lithium ion conduction in PEO–salt electrolytes gelled with PAN. *Solid State Ionics* **113**, 123–127.

10 http://www.sonyenergy-devices.co.jp/en/csr/quality.php.

11 Xu, J.J., Ye, H., Huang, J. (2005) Novel zinc ion conducting polymer gel electrolytes based on ionic liquids. *Electrochemistry Comm.* **7**, 1309–1317.

12 Scrosati, B. (2002) Lithium polymer electrolytes. In W. Van Schalkwijk, B. Scrosati (eds) *Advances in Lithium-ion batteries*, Kluwer Academic Publishers, New York, NY, 251–267.

13 Kato, Y., Hori, S., Saito, T., Suzuki, K., Hirayama, M., Mitsui M. *et al.* (2016) High-power all-solid-state batteries using sulfide superionic conductors. *Nature Energy* **1**, 16030.

14 Narayanan, R., Aswin Manohar, K., Ratnakumar, B.V. (2015) Fundamental aspects of ion transport in solid electrolytes. In N.J. Dudney, W.C. West, J. Nanda (eds) *Handbook of Solid State Batteries*, 2nd edn, World Scientific, Singapore, 3–50.

15 http://www.infinitepowersolutions.com/.

16 Tehrani, Z., Korochkina, T., Govindarajan, S., Thomas, D.J., O'Mahony, J., Kettle, J. *et al.* (2015) Ultra-thin flexible screen printed rechargeable polymer battery for wearable electronic applications. *Organic Electronics* **26**, 386–394.

17 Kim, S.H., Choi, K.H., Cho, S.J., Choi, S., Park, S., Lee, S.Y. (2015) Printable solid-state lithium-ion batteries: a new route toward shape-conformable power sources with aesthetic versatility for flexible electronics. *Nano Lett.* **15**, 5168–5177.

18 Steingart, D., Ho, C.C., Salminen, J., Evans, J.W., Wright, P.K. (2007) Dispenser printing of solid polymer-ionic liquid electrolytes for lithium ion cells. In *Polytronic 2007–6th International Conference on Polymers and Adhesives in Microelectronics and Photonics*, IEEE, 261–264.

19 Wang, Y., Liu, B., Li, Q., Cartmell, S., Ferrara, S., Deng, Z.D. *et al.* (2015) Lithium and lithium ion batteries for applications in microelectronic devices: a review. *J. Power Sources* **286**, 330–345.

20 Gaikwad, A.M., Arias, A.C., Steingart, D.A. (2015) Recent progress on printed flexible batteries: mechanical challenges, printing technologies, and future prospects. *Energy Technology* **3**, 305–328.

21 Kim, B., Winslow, R., Lin, I., Gururangan, K., Evans, J., Wright, P. (2015) Layer-by-layer fully printed Zn-MnO2 batteries with improved internal resistance and cycle life. *J. Physics: Conference Series* **660(1)**, 012009, IOP Publishing.

22 Singh, N., Galande, C., Miranda, A., Mathkar, A., Gao, W., Reddy, A.L.M. *et al.* (2012) Paintable battery. *Scientific Reports*, 2.

23 Sousa, R.E., Costa, C.M., Lanceros-Méndez, S. (2015) Advances and future challenges in printed batteries. *ChemSusChem* **8**, 3539–3555.

24 Hu, Y., Sun, X. (2014) Flexible rechargeable lithium ion batteries: advances and challenges in materials and process technologies. *J. Mater. Chem. A* **2**, 10712–10738.

25 Stephan, A.M. (2006) Review on gel polymer electrolytes for lithium batteries. *European Polymer J.* **42(1)**, 21–42.

26 Hon, K.K.B., Li, L., Hutchings, I.M. (2008) Direct writing technology—advances and developments. *CIRP Annals–Manufacturing Technology* **57**, 601–620.

27 Ferrari, S., Loveridge, M., Beattie, S.D., Jahn, M., Dashwood, R.J., Bhagat, R. (2015) Latest advances in the manufacturing of 3D rechargeable lithium microbatteries. *J. Power Sources* **286**, 25–46.

28 Gross, B.C., Erkal, J.L., Lockwood, S.Y., Chen, C., Spence, D.M. (2014) Evaluation of 3D printing and its potential impact on biotechnology and the chemical sciences. *Anal. Chem.* **86**, 3240–3253.

29 Singh, M., Haverinen, H.M., Dhagat, P., Jabbour, G.E. (2010) Inkjet printing—process and its applications. *Adv. Mater.* **22**, 673–685.

30 Ho, C.C., Evans, J.W., Wright, P.K. (2010) Direct write dispenser printing of a zinc microbattery with an ionic liquid gel electrolyte. *J. Micromech. Microeng.* **20(10)**, 104009.

31 Wright, P.K., Dornfeld, D.A., Chen, A., Ho, C.C., Evans, J.W. (2010) Dispenser printing for prototyping microscale devices. *Transactions of NAMRI/SME* **38**.

32 Ho, C.C., Evans, J.W., Wright, P.K. (2009) Direct write dispenser printing of zinc microbatteries. In *Proceedings of the 9th Power MEMS Workshop*.

33 Meyer, W.H. (1998) Polymer electrolytes for lithium-ion batteries. *Adv. Mater.* **10**, 439–448.

34 Dudney, N.J. (2005) Solid-state thin-film rechargeable batteries. *Mater. Sci. Eng. B* **116(3)**, 245–249.

35 MacKenzie, J.D., Ho, C. (2015) Perspectives on energy storage for flexible electronic systems. *Proc IEEE* **103**, 535–553.

36 Zhou, G., Li, F., Cheng, H.M. (2014) Progress in flexible lithium batteries and future prospects. *Energy Environ. Sci.* **7**, 1307–1338.

37 Braam, K.T., Volkman, S.K., Subramanian, V. (2012) Characterization and optimization of a printed, primary silver–zinc battery. *J. Power Sources* **199**, 367–372.

38 Hilder, M., Winther-Jensen, B., Clark, N.B. (2009) Paper-based, printed zinc–air battery. *J. Power Sources* **194**, 1135–1141.

39 Wendler, M., Hübner, G., Krebs, M. (2011) Development of printed thin and flexible batteries. *Sci. Technol.* **4**, 32–41.

40 Delannoy, P.E., Riou, B., Lestriez, B., Guyomard, D., Brousse, T., Le Bideau, J. (2015) Toward fast and cost-effective ink-jet printing of solid electrolyte for lithium microbatteries. *J. Power Sources* **274**, 1085–1090.

41 Guyomard-Lack, A., Delannoy, P.-E., Dupre, N., Cerclier, C.V., Humbert, B., Le Bideau, J. (2014) Destructuring ionic liquids in ionogels: enhanced fragility for solid devices. *Phys. Chem. Chem. Phys.* **16**, 23639.

42 www.InfinitePowerSolutions.com.

43 www.ilika.com.

44 Fu, K., Wang, Y., Yan, C., Yao, Y., Chen, Y., Dai, J. *et al.* (2016) Graphene oxide-based electrode inks for 3D-printed lithium-ion batteries. *Adv. Mater.* **28**, 2587–2594.

45 Selimis, A., Mironov, V., Farsari, M. (2015) Direct laser writing: principles and materials for scaffold 3D printing. *Microelectronic Eng.* **132**, 83–89.

46 Deckard, C. (2010) Method of selective laser sintering with improved materials. US 7794647 B1.

47 Das, S., Beama, J.J., Wohlert, M., Bourell, D.L. (1995) Direct laser freeform fabrication of high performance metal components. *Rapid Prototyping J.* **4(3)**, 112–117.

48 Lewis, J.A., Gratson, G.M. (2004) Direct writing in three dimensions. *Materials Today* **7(7)**, 32–39.

49 Kim, H., Sutto, T.E., Piqué, A. (2014) Laser materials processing for micropower source applications: a review. *J. Photonics Ener.* **4(1)**, 040992–040992.

50 Arnold, C.B., Serra, P., Piqué, A. (2007) Laser direct-write techniques for printing of complex materials. *MRS Bull.* **32(01)**, 23–31.

51 Ollinger, M., Kim, H., Sutto, T., Martin, F., Piqué, A. (2006) Laser direct-write of polymer nanocomposites. *J. Laser Micro. Nanoeng.* **1(2)**, 102–105.

52 Ollinger, M., Kim, H., Sutto, T., Piqué, A. (2006) Laser printing of nanocomposite solid-state electrolyte membranes for Li micro-batteries. *Appl. Surf. Sci.* **252**, 8212–8216.

53 Kim, H., Sutto, T.E., Proell, J., Kohler, R., Pfleging, W., Piqué, A. (2014) Laser-printed/structured thick-film electrodes for Li-ion microbatteries. SPIE LASE 89680 L-89680 L, International Society for Optics and Photonics.

54 Sutto, T.E., Ollinger, M., Kim, H., Arnold, C.B., Piqué, A. (2006) Laser transferable polymer-ionic liquid separator/electrolytes for solid-state rechargeable lithium-ion microbatteries. *Electrochem. Solid-State Lett.* **9(2)**, A69–A71.

55 Lan, H., Ding, Y. (2010) Nanoimprint lithography. In M. Wang (ed) *Lithography*, http://www.intechopen.com/books/lithography/nanoimprint-lithography.

56 Kil, E.H., Choi, K.H., Ha, H.J., Xu, S., Rogers, J.A., Kim, M.R. *et al.* (2013) Imprintable, bendable, and shape-conformable polymer electrolytes for versatile-shaped lithium-ion batteries. *Adv. Mater.* **25**, 1395–1400.

5

Design of Printed Batteries: From Chemistry to Aesthetics

Keun-Ho Choi and Sang-Young Lee

Department of Energy Engineering, School of Energy and Chemical Engineering, Ulsan National Institute of Science and Technology (UNIST), Korea

5.1 Introduction

With the advent of flexible/wearable electronics and the Internet of Things (IoT), which are expected to drastically change our daily lives, printed electronics have drawn much attention as a low-cost, efficient, and scalable platform technology [1–3]. Printed electronics require so-called "printed batteries" as a monolithically integrated power source that can be prepared by the same printing processes (Figure 5.1).

From the battery manufacturing point of view, conventional batteries with fixed shapes and sizes are generally fabricated by winding or stacking cell components (such as anodes, cathodes, and separator membranes) and then packaging them with (cylindrical-/rectangular-shaped) metallic canisters or pouch films, finally followed by injection of liquid electrolytes [3, 4]. In particular, the use of liquid electrolytes gives rise to serious concerns in cell assembly because they require strict packaging materials to avoid leakage problems and also separator membranes to prevent electrical contact between electrodes. For these reasons, conventional battery materials and assembly processes have caused lack of variety in form factors, thus imposing formidable challenges in relation to their integration into versatile-shaped electronic devices. In contrast to conventional batteries, printed batteries are thinner and lighter, and can be easily produced through a variety of cost-effective printing processes [4]. More importantly, they offer unprecedented opportunities to diversify battery

Printed Batteries: Materials, Technologies and Applications, First Edition.
Edited by Senentxu Lanceros-Méndez and Carlos Miguel Costa.
© 2018 John Wiley & Sons Ltd. Published 2018 by John Wiley & Sons Ltd.

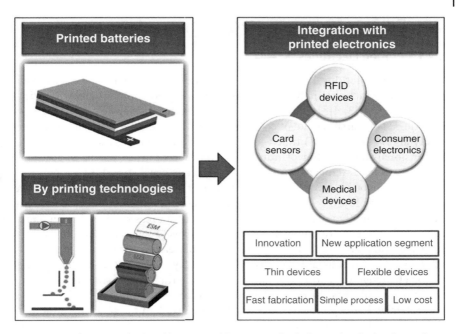

Figure 5.1 Schematic of printed battery architecture and printing technologies. Examples and features of printed battery electronics integration are also described [4].

design, dimensions, and form, far beyond that achievable with conventional battery technologies.

Printing technology is a facile and reproducible process in which slurries or inks are deposited to make pre-defined patterns [4, 5]. The slurries/inks should be designed to fulfill the requirements (such as rheology and particle dispersion) of the printing process. The development of printed batteries involves the design and fabrication of battery component slurries/inks. Most studies of printed batteries have been devoted to the development of printed electrodes. However, in order to reach the ultimate goal of so-called "all-printed batteries", printed separator membranes and electrolytes should be also developed along with printed electrodes.

In this chapter, we describe the design principle and recent advances in printed batteries, with a focus on the major components. Based on this understanding, the current status of and research progress in various printed battery systems, including Zn/MnO_2 batteries, supercapacitors, lithium-ion batteries, and other batteries, are presented with particular attention to their aesthetic versatility. In addition, development direction and future prospects as a potential power source with design flexibility are discussed.

5.2 Design of Printed Battery Components

5.2.1 Printed Electrodes

A primary step for developing printed electrodes is the preparation of electrode slurry/inks, which should be tailored to ensure process compatibility with printing techniques [3, 4]. Key requirements for the printed electrode slurry/inks include: (1) dispersion state of components, (2) rheological properties (i.e., viscosity and viscoelasticity) tuned for specific printing processes, (3) structural/dimensional stability after printing (e.g., cohesion between particles, adhesion with substrates and mechanical tolerance upon external stress) [6].

Among various electrode components, electrically conductive additives are essentially employed to facilitate electron transport in electrodes. Of readily available conductive additives, carbonaceous substances such as carbon nanotubes (CNTs) and graphenes have been extensively investigated due to their high electrical conductivity and also affordable electrochemical capacitance, which is often beneficial for electrical double-layer capacitors (EDLCs) [7]. A major challenge for carbonaceous substances is their dispersion in water and other polar solvents. In particular, for CNT cases, strong inter-tube affinity and intrinsic hydrophobicity pose a stringent problem for securing good dispersion state [7, 8]. In most studies, the dispersion of CNTs has been achieved with the assistance of CNT-dispersing agents such as SDBS (sodium dodecylbenzene sulfonate), and surface functionalization of CNTs [7, 8]. An important note is that these additional procedures for CNT dispersion should not impair intrinsic electrical conductivity of CNTs.

The electrically conductive additives are mixed with electrochemically active materials in the presence of binders, which are used to allow cohesion between the electrode components and also adhesion with current collectors. To develop printed electrodes with various form factors and reliable physico-chemical/electrochemical properties, control of the rheological properties of electrode slurries/inks is needed as an important prerequisite, along with rational design of the electrode chemistry. Details of the theoretical understanding of the rheological behavior of slurries and inks were described in review papers [3, 9, 10], with a particular focus on the relationship with coating and printing techniques. The materials, compositions, and rheology of electrode slurry/inks should be fine-tuned to meet the requirements of the chosen printing process. Information on materials and formulations of various electrode slurries/inks, which is classified as a function of battery types, is provided in Table 5.1.

To enable electrodes to be electrochemically active in batteries, ionic pathways as well as electronic channels in electrodes should be simultaneously secured. One interesting point in Table 5.1 is that most of the electrode slurries/inks contain only electrically conductive components (represented by

Table 5.1 Materials and formulations of electrode slurries/inks for various printed batteries.

Battery type	Active materials	Conducting agents	Binding agents	Solvents	Additives	Printing technique	Ref
Supercapacitors	Activated carbon (3 wt%)	–	Polytetrafluoroethylene (PTFE) (5 wt%)	Ethylene glycol	Triton X100	Inkjet printing	[11]
LIBs	$Li_4Ti_5O_{12}$ (1.3 wt%), $LiFePO_4$ (1.5 wt%)	–	–	Ethylene glycol (20–30 wt%) DI water (31.7 ~ 38.5 wt%)	Glycerol (20 ~ 27 wt%) Aqueous hydroxypropyl cellulose (HPC) (8 ~ 9 wt%) Aqueous hydroxyethyl cellulose (HEC) (1 ~ 2 wt%)	3D printing	[12]
LIBs	$LiFePO_4$ (8 wt%)	Carbon black (1 wt%)	Poly-acrylic-comaleic acid copolymer (0.5 wt%)	DI water (90 wt%)	Triton X100	Inkjet printing	[13]
Li-S	Aligned multiwall carbon nanotubes (MWCNT) (0.023 wt%)	–	–	DI water (92.691 wt%)	Isopropyl alcohol (7.286 wt%)	Dispenser printing	[14]
Li-O$_2$	Carbon black (10.7 wt%)	–	Nafion (2.6 wt%)	2-propanol (86.7 wt%)	–	Screen printing	[15]
LIBs	Graphite (38.8 wt%)	–	Carboxymethylcellulose (CMC) (0.8 wt%)	DI water (60 wt%)	Microfibrillated cellulose (MFC) (0.4 wt%)	Screen printing	[16]

(Continued)

Table 5.1 (Continued)

Battery type	Active materials	Conducting agents	Binding agents	Solvents	Additives	Printing technique	Ref
LIBs	$LiFePO_4$	Carbon black	CMC	DI water	HCl, NaOH, Triton X-100, Glycerin	Inkjet printing	[17]
LIBs	$LiFePO_4$ (24.12 wt%)	Carbon black (3.015 wt%)	poly(vinylidene fluoride) (PVDF) (3.015 wt%)	N-methyl-2-pyrrolidone (NMP) (69.85 wt%)		Screen printing	[18]
LIBs	$Li_4Ti_5O_{12}$ or $LiMn_2O_4$ (60 wt%)	Ketjen black (6 wt%)	Poly(ethylene glycol) (PEG) (26 wt%)	–	LiTFSI (8 wt%)	Screen printing	[19]
LIBs	Zr-incorporated $LiCoO_2$ (95 wt%)	Ag powders (5 wt%)	–	2-methoxy ethanol, acetic acid	–	Screen printing	[20]
LIBs	SnO_2	Acetylene black	CMC	DI Water, Ethanol, Diethylene glycol, Triethanolamine, Isopropylalcohol	CH10B, CH12B	Inkjet printing	[21]
LIBs	$LiCoO_2$	Carbon black	Ethyl cellulose	Terpineol	PS-21A	Screen printing	[22]
LIBs	$LiCoO_2$	Carbon black, Graphite	Ethyl cellulose, Epoxy resin	Terpineol Butyl glycidyl ether Butyl cellulose Mineral spirit	PS-21A	Screen printing	[23]

Application	Active material	Conductive additive	Binder	Solvent	Additive	Method	Ref.
LIBs	$LiCoO_2$ (2.64 wt%)	Carbon black (0.13 wt%)	CMC (0.013 wt%)	DI water (88.13 wt%)	Lomar D, Monoethanolamine (9.08 wt%)	Inkjet printing	[24]
LIBs	Graphene oxide (16 wt%)	–	Poly(sodium 4-styrenesulfonate), or poly[2,5-bis(3-sulfonatopropoxy)-1,4-ethynylphenylene-alt-1,4-ethynylphenylene] sodium salt (60 wt%)	DI water (24 wt%)	–	Drop casting	[25]
LIBs	$Li_4Ti_5O_{12}$ (40 wt%)	Acetylene black (5 wt%)	PVdF (5 wt%)	NMP (50 wt%)	–	Dispenser printing	[26]
LIBs	$LiCoO_2$ Natural graphite	Carbon black Carbon black	PVdF, SBR/CMC	NMP DI water	–	Screen printing	[27]
LIBs	$Li_4Ti_5O_{12}$ (40 wt%)	Acetylene black (5 wt%)	PVdF (5 wt%)	NMP (50 wt%)	–	Doctor blade	[28]
$Ni-H_2$	β-Nickel hydroxide powder	Zinc, Cobalt (5.2 wt%)	Polyacrylic acid potassium salt (PAAK)	KOH solution (26 wt%)	–	Thick film printing	[29]
LIBs	$Li_4Ti_5O_{12}$ $LiCoO_2$	Carbon black	PVdF	NMP	–	Dip coating	[30]
LIBs	$Li_4Ti_5O_{12}$ $LiCoO_2$	Carbon black	PVdF	NMP	–	Doctor blade	[31]

(*Continued*)

Table 5.1 (Continued)

Battery type	Active materials	Conducting agents	Binding agents	Solvents	Additives	Printing technique	Ref
Zn/MnO$_2$	Zn powders (69.3 wt%)	–	Styrene-butadiene (1.6 wt%)	Ethylene glycol (10.9 wt%)	ZnO nanopowder (7.3 wt%), Bi$_2$O$_3$ (10.9 wt%)	Stencil printing	[32]
	A: Zn powders (75~77 wt%) C: MnO$_2$ (48~53 wt%)	C: Graphite (9~13 wt%)	A: PEO (1.3~1.5 wt%) C: PEO (1.4~1.8 wt%)	H$_2$O (21~24 wt%) (37~39 wt%)	–	Screen printing	[33]
	MnO$_2$ powder (9 g)	Carbon black (0.5 g)	SBR (6.6 g)	DI water (4 g)	–		[34]
	MnO$_2$ powder (90 wt%)	Acetylene black (6 wt%)	PSBR (4 wt%)	DI water	–	Flexographic printing	[35]
Zn/Ag	Zn:Ag$_2$O (3~4:1)	Acetylene black	PVA (1 wt%), PVDF (1 wt%)	NMP	–	Extrusion printing	[36]
	Zn		Methylcellulose (5 wt%)	DI water	–	Stencil printing	[37]
	Ag$_2$O Silver		PEO	DI water n-tetradecane	–	Inkjet printing	[38]
Zn/Air	Zn powders (88 wt%)	CNF (3 wt%), CB (1 wt%)	PVDF-HFP (8 wt%)	Acetone	–	Flexographic printing	[39]
	Co$_3$O$_4$ nanoparticles (1 mg mL-1)	Carbon black	Anionic polymer binder (AS-4)	Acetone			
	Zn powder	Carbon powder	Polycarbonate (15 wt%)	Tetrahydrofuran	–	Screen printing	[40]

carbon-based additives) and not ionically conductive ones. As a consequence, the resultant printed electrodes should uptake liquid electrolytes after being assembled with separator membranes and packaging substances. Recently, Lee *et al.* demonstrated a new electrode slurry that afforded mixed electron/ion conduction (Figure 5.2) [1]. The electrode slurries were composed of electrochemically active powders (such as LiFePO$_4$ (cathode) and Li$_4$Ti$_5$O$_{12}$ (anode)), carbon black conductive additives, and ionically conductive matrix (= ultraviolet (UV)-crosslinkable ethoxylated trimethylolpropane triacrylate (ETPTA) monomer + 1 M LiPF$_6$ in ethylene carbonate (EC)/propylene carbonate (PC) as

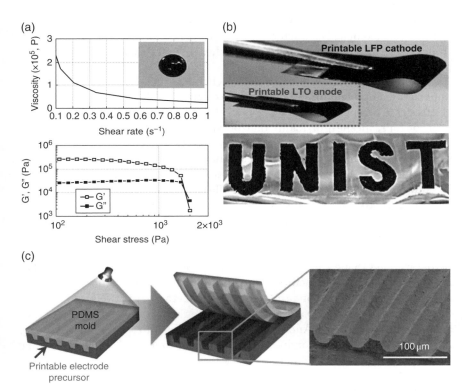

Figure 5.2 Printed electrodes with mixed electron/ion conduction. (a) Rheological properties (viscosity and viscoelasticity represented by storage modulus (G′) and loss modulus (G″)) of the electrode slurry, which consisted of LiFePO$_4$ (as cathode material), or Li$_4$Ti$_5$O$_{12}$ (as anode material), carbon black additives, and ionically conductive matrix (= UV-crosslinkable ETPTA monomer + 1 M LiPF$_6$ in EC/PC) without NMP solvent. (b) Photographs showing mechanical flexibility (upper image) and printability (represented by word "UNIST", lower image) of the electrode slurry. (c) Schematic illustration of the UV-IL technique-driven micropatterning procedure and an SEM image showing the printed electrode with an inverse replica of the finely defined microscale stripe pattern. Reproduced with permission [1].

a thermally tolerant electrolyte). In comparison to the previously reported systems, these printed electrodes possessed the electrolyte-conductive as well as the carbon-conductive additives, resulting in the elimination of liquid-electrolyte injection steps and also traditional processing solvents such as water and NMP (N-methyl-2-Pyrrolidone) that demand additional time-/energy-consuming drying processes. The rheological properties of the electrode slurries were controlled by adjusting the composition ratio and solid content to show thixotropic fluid behavior [41, 42] that is suitable for a stencil-printing process. The electrode inks were conformally printed onto various substrates (including curvilinear surfaces) by stencil printing. Subsequently, the printed electrode inks were solidified after exposure to UV irradiation in less than a minute. Driven by such well-tuned rheological characteristics and process simplicity, the printed electrodes with various form factors (such as heart-/letter-shapes and microscale patterns) were successfully realized.

Current collectors, along with electrochemically active materials, are an important component for fast and reliable electrochemical reactions in battery electrodes. Current collectors serve as an electrically conductive support in the form of foils, meshes, or foams that provide pathways for efficient electron transport towards electrochemically active layers [3]. The current collectors should be chemically compatible and also have good adhesion with the electrochemically active layers to maintain the structural integrity of the electrodes. In addition, a wide electrochemical-stability window is required to prevent unwanted corrosion reaction during battery operation. To develop printed electrodes with design versatility, the current collectors should be also printable in addition to the other printed battery components. Currently, the usual current collectors in commercial electrodes are metallic foils; however, their predetermined shape and mechanical stiffness pose impediments to diversifying the design and form factors of the printed batteries. Unfortunately, only a few works have reported on printed current collectors. Cui *et al.* presented the CNT-coated conductive paper current collector, which showed a sheet resistance of ~ 1 ohm sq^{-1} [43]. Another approach was the utilization of a spray-coating method to produce a conductive CNT current collector. The spray-coating solution consisted of CNT/carbon black (=8/2 (w/w)) in NMP solvent [44]. The sheet resistance of the resulting spray-coated CNT current collector was found to be as low as ~ 10 ohm sq^{-1}. In addition to the CNT current collector, a spray-coated metal current collector was prepared using Cu (copper) inks [44]. This result underscores the potential applicability of metallic inks as a printed current collector for use in printed batteries. A variety of commercially available metallic inks which can be used in printed batteries are summarized in Table 5.2.

Substrates, onto which the electrode components are printed, also play a valuable role in achieving high-precision printed batteries. For example, printing on so-called "wetting substrates", such as conventional paper or textiles,

Table 5.2 Commercially available metallic inks.

Materials	Curing condition	Resistivity/sheet resistance	Manufacturer
Cu	160 °C for 60 min	1.0×10^{-4} ohm cm^{-1} /N. A.	Tatsuta Electric Wire & Cable Co., Ltd
Ni	93 °C for 120 min	N. A. /2.0–3.5 ohm sq^{-1}	TED PELLA. INC.
Al	170–270 °C for 10 min	Low resistivity /N. A.	Dupont
Ag	120–150 °C for 30–60 min	1.0–3.0×10^{-5} ohm cm^{-1} /1.59×10^{-6} ohm cm^{-1}	Aldrich, Dupont

tends to provoke random spreading of ink droplets due to uncontrollable capillary force created by the inhomogeneously distributed/several-hundred-micrometer-sized pores of the substrates [45, 46]. Meanwhile, "non-wetting substrates", such as non-porous polyethylene terephthalate (PET) films, often result in a ring-like deposition (i.e., coffee-ring formation) of ink particles [45] as well as having poor adhesion between the printed products and the substrates [6]. To resolve substrate-triggered printing failure, previous work has been mostly devoted to tuning ink materials and formulations, which often limits the selection of ink chemistry. As a facile and efficient approach to addressing this substrate issue, Lee *et al.* demonstrated an inkjet-printed cellulose nanomat on commercial A4 paper (Figure 5.3) [2]. The cellulose nanomat layers were designed to exhibit a nanoporous structure and uniform surface roughness, eventually providing the combined advantageous attributes of "wetting" and "non-wetting" substrates. As a consequence, the cellulose nanomat layer contributed to broadening the chemistry/formulation spectrum of electrode/electrolyte inks and also to achieving high-resolution print images on conventional A4 paper without the concerns of droplet spreading or coffee-ring formation.

5.2.2 Printed Separator Membranes and Solid-state Electrolytes

Conventional rechargeable batteries have (cylindrical-/rectangular-shaped) stereotypical forms which are generally fabricated using a winding or stacking process. Such cell assembly has meant that power sources lack a variety of form factors and design flexibility. This unwanted limitation is believed to mostly arise from the use of separator membranes and liquid electrolytes. In most of the printed batteries investigated so far, separator membranes and electrolytes have not been a center of attention, as compared to electrodes. However, in

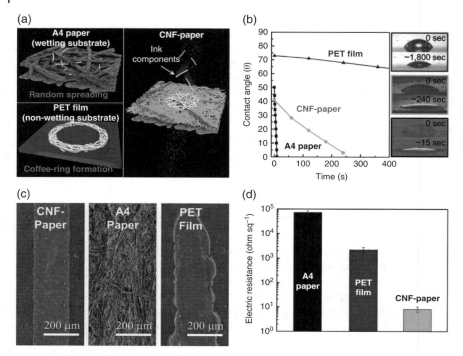

Figure 5.3 Inkjet-printed cellulose nanomat on commercial A4 paper. (a) Effect of substrates on the resolution of the inkjet printing process: the wetting substrate (left upper side, random spreading of ink droplets), the non-wetting substrate (left lower side, coffee-ring formation), and the CNF nanomat on A4 paper (right side, high-resolution print pattern). (b) Variation in the water-contact angles of different substrates with time. (c) SEM images (surface view) of the inkjet-printed ((SWNT/AC) + Ag NWs) electrodes on different substrates. (d) Electric resistances of the inkjet-printed ((SWNT/AC) + Ag NWs) electrodes on different substrates. Reproduced with permission [2]. (*See insert for color representation of the figure.*)

order to reach the ultimate goal of so-called "all-printed-batteries", printed separator membranes and printed electrolytes should be developed along with the aforementioned printed electrodes.

Subramanian *et al.* developed a printed, UV-cured gel separator membrane [37] (Figure 5.4), which consisted of crosslinked polyacrylic acid and poly(ethylene oxide) (PEO) fillers that improve the mechanical stability of the resulting separator membrane and control rheological properties. The crosslinked polyacrylic acid was synthesized by a free-radical method using a water-soluble photo initiator. The crosslinking density of the polyacrylic acid was controlled to optimize the mechanical strength and the water-swelling ratio. The 1 wt% crosslinking density was found to give optimal performance overall. Lower concentrations resulted in incomplete gel formation with

(a)   (b)

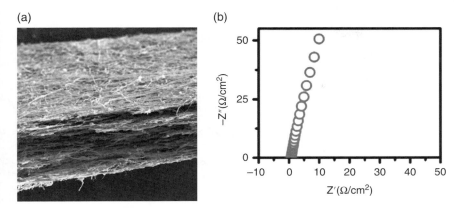

Figure 5.4 Printed, UV-cured gel separator membrane. (a) Polymerization scheme of polyacrylic gel electrolyte, where R is the water-soluble photoinitiator, 4-(2-hydroxyethoxy) phenyl(2-hydroxyl-2-propyl) ketone. (b) Photolysis of the photoinitiator after exposure to UV light. (c) Photograph of the polymerized gel separator. Reproduced with permission [37].

insufficient mechanical strength. On the other hand, higher concentrations $(4 \sim 5\,wt\%)$ showed poor solubility, causing deterioration of printing processability and the properties of the printed electrolytes.

Ajayan *et al.* fabricated a spray-printed separator membrane (Figure 5.5) [44]. The printing solution was composed of polyvinylidene fluoride-hexafluoro-propylene (PVdF-HFP)/polymethylmethacrylate (PMMA)/and fumed $SiO_2 =$ 27/9/4 (w/w/w) in a solvent mixture of acetone/N,N-dimenthylformamide (DMF) = 1/1 (v/v). PVdF-HFP, which has good solubility in low-boiling-point solvents, acted as a polymer matrix to endow mechanical rigidity, while PMMA was chosen to promote adhesion towards substrates. The resulting spray-printed separator membranes showed a fibrous morphology with high porosity, which was favorable for absorbing liquid electrolyte.

Lee *et al.* demonstrated a printed, flexible solid-state electrolyte (Figure 5.6) [47, 48]. The printed solid-state electrolyte consisted of UV-crosslinked ETPTA polymer network as a mechanical framework, EC/PC-based high-boiling point electrolyte, and Al_2O_3 nanoparticles as a nano spacer/viscosity-controlling agent. The rheological behavior of the printable electrolyte pastes was affected by the composition ratio and dispersion state of Al_2O_3 nanoparticles. At the optimum composition ratio, the printable electrolyte pastes exhibited shear-thinning behavior (i.e., decrease of viscosity with shear rate) and also crossover of elastic/loss modulus at a certain shear stress, revealing thixotropic fluid behavior. In contrast, conventional liquid electrolytes showed newtonian fluid-like rheological behavior. The electrolyte pastes with well-tuned rheological characteristics were printed onto various substrates using a stencil-printing process. After exposure to UV irradiation, the as-printed electrolyte pastes

(a)

(b)

(c)

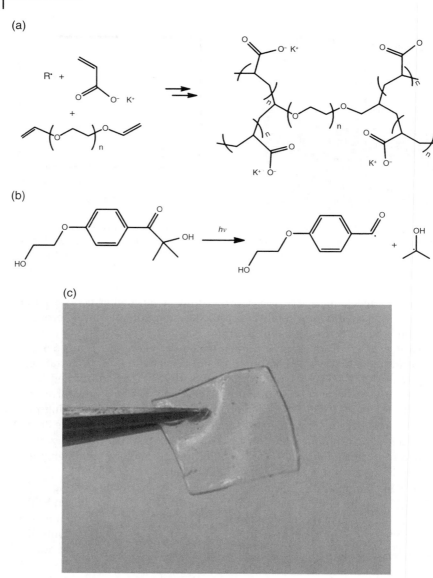

Figure 5.5 Spray-printed separator membranes. (a) SEM image showing the layered and fibrous structure. (b) Nyquist plot of the spray-printed separator membrane. The separator shows an ionic conductivity of ~ 1.24 mS cm^{-1}. Reproduced with permission [44].

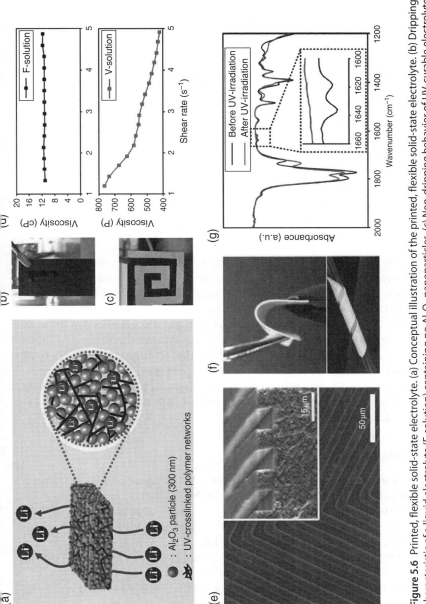

Figure 5.6 Printed, flexible solid-state electrolyte. (a) Conceptual illustration of the printed, flexible solid-state electrolyte. (b) Dripping characteristic of a liquid electrolyte (F-solution) containing no Al_2O_3 nanoparticles. (c) Non-dripping behavior of UV-curable electrolyte mixture (V-solution) containing Al_2O_3 nanoparticles before UV-crosslinking reaction. (d) Comparison of viscosity between the F- and V-solution as a function of shear rate. (e) SEM images (surface) of the printed solid-state electrolyte. (f) Photographs showing the highly bendable and twistable features of the printed solid-state electrolyte. (g) FT-IR spectra depicting acrylic C=C double bonds of the printed solid-state electrolyte before/after UV irradiation. Reproduced with permission [47].

were solidified. Notably, the printable electrolyte pastes, in combination with the UV-IL (imprint lithography) technique, enabled the fabrication of microscale-structured patterns. Similar approaches were applied to the development of ionic liquid-based solid-state electrolytes. The ionic liquid was mixed with the UV-curable acrylate monomer [2] or confined in silica matrix via sol-gel reaction [49], leading to printable solid-state electrolytes.

5.3 Aesthetic Versatility of Printed Battery Systems

Most of the printed batteries reported to date were devoted to zinc (Zn)-based electrochemistry. Exploration of other battery systems such as supercapacitors and lithium-ion batteries for potential use as a printed power source has been recently undertaken. The types, components, printing techniques, and electrochemical characteristics of the printed battery systems reported to date are summarized in Table 5.3.

Besides the materials and associated electrochemistry, the design and configuration of cell components are also important in developing high-performance printed batteries with aesthetic versatility. In commercial batteries, anodes and cathodes are stacked in series with separator membranes (Figure 5.7(a)). This in-series configuration is beneficial in reducing the distance between the anode and cathode, thereby lowering internal ohmic resistance of the cell. In addition to the in-series configuration, in-parallel model can be applied to batteries, wherein the anode and the cathode are positioned side-by-side (Figure 5.7(b)). The in-parallel configuration is effective for widening the printing process window owing to the process simplicity and minimal risk of short-circuit failure. Printed batteries require a broad variety of form factors and shape deformability. In this respect, the in-parallel configuration could be a promising way to enrich the design and applicability of printed batteries. Below is an overview of various printed batteries in terms of electrochemical cell system and design diversity.

5.3.1 Zn/MnO$_2$ Batteries

The R2R (roll-to-roll) processable Zn/MnO$_2$ battery was fabricated using a fluidic cathode paste [53]. The prototype Zn/MnO$_2$ battery consisted of a Zn foil as an anode and current collector, a separator membrane soaked with NH$_4$Cl and ZnCl$_2$ electrolyte solution, and a solution-processable MnO$_2$ paste as a cathode. The MnO$_2$ paste contained MnO$_2$ active materials, conductive CNT networks, and electrolyte solution, wherein the CNTs acted as a current collector and also conductive agent.

Steingart *et al.* reported printed flexible Zn/MnO$_2$ batteries based on Nylon mesh-containing electrodes. This new structure was effective in reducing the

Table 5.3 Types, components, printing techniques, and electrochemical characteristics of various printed power sources.

Battery type	Cathode	Anode	Separator and electrolyte	Capacity/voltage	Printing technique	References
Supercapacitors	Activated Carbon/SWNT	Activated Carbon/SWNT	ETPTA/ [BMIM][BF_4]	100 mF cm^{-2} /2.0 V	Inkjet printing	[2]
Polymer Battery	PEDOT:PSS	PEDOT:PSS / PEI	PSSNa	5.5 mAh g^{-1}	Screen printing	[50]
LIB	$LiFePO_4$ (LFP)	$Li_4Ti_5O_{12}$ (LTO)	PVDF-co-HFP and Al_2O_3	100 mAh g^{-1} /2.5 V	3D printing	[51]
	$LiFePO_4$ (LFP)	$Li_4Ti_5O_{12}$ (LTO)	ETPTA/ 1M $LiPF_6$ in EC/PC (1:1 v/v) mand Al_2O_3	160 mAh g^{-1} /2.5 V	Stencil printing	[1]
Zn/MnO_2	MnO_2	Zn	PAA-based gel electrolyte	0.8 mAh cm^{-2} /1.5 V	Stencil printing	[32]
	MnO_2	Zn	37% KOH with 3% ZnO	0.4 mAh cm^{-2} /1.4 V	Screen printing	[33]
	MnO_2	Zn	zinc triuoromethanesulfonate (ZnOtf) (0.2 g) 1-butyl-3-methylimidazolium triuoromethanesulfonate ([BMIM][Otf]) (3 g)	0.05 mAh cm^{-2} /1.3 V	Doctor blade casting	[34]
	MnO_2	Zn	1:1 = PVDF-HFP: 0.5 M solution of zinc trifluoromethanesulfonate (in BMIM + Tf-)	1.0 mAh cm^{-2} /1.6 ~ 1.8 V	Flexographic printing	[35]

(Continued)

Table 5.3 (Continued)

Battery type	Cathode	Anode	Separator and electrolyte	Capacity/voltage	Printing technique	References
Zn/Ag	Ag	Zn	PAA-based gel 6 M KOH + 1 M LiOH	2.1 mA h cm^{-2} /1.5 V	Screen printing	[52]
	Ag$_2$O	Zn	Methylcellulose 17 M KOH (57:29:14 = H$_2$O:KOH:PEO)	3.95 mWh cm^{-2} /1.4 V	Extrusion printing	[36]
	Ag$_2$O	Zn	PAA-based gel electrolyte 8.4 M KOH	5.4 mA h cm^{-2} /1.55 V	Stencil printing	[37]
	Ag	Zn	10 M KOH	3.95 mWh cm^{-2} /1.5, 1.8 V	Inkjet printing	[38]
Zn/Air	Co$_3$O$_4$	Zn	PVA-gelled electrolyte membrane (KOH/PVA = 35/2)	2905 Wh L^{-1} /1.2~1.3 V	Doctor blade casting	[39]
	PEDOT	Zn	8 M LiCl/LiOH	0.5 mAh cm^{-2} /0.8 V	Screen printing	[40]

Figure 5.7 Schematic illustration of batteries with different component configurations: (a) in series (b) in parallel.

(a)

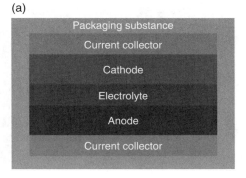

(b)

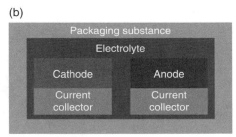

overall electrode thickness and improving the capacity without impairing the power density (Figure 5.8) [54]. The Nylon mesh-embedded electrodes were fabricated using a stencil-printing process. The Zn and MnO_2 slurries were sequentially stencil-printed, followed by printing of a silver ink as a current collector. Due to the incorporation of the Nylon mesh serving as a mechanical scaffold, the resulting mesh-embedded electrodes showed significantly higher thicknesses (>300 μm) compared to those (60 ~ 75 μm) of control electrodes incorporating no Nylon mesh, thereby enabling a significant increase in the areal capacity of the electrodes. Finally, the battery was assembled by stacking the mesh-embedded electrodes and polyacrylic acid gel polymer electrolyte-soaked Nylon mesh separator. The resulting battery provided a discharge capacity of 4.5 mAh cm^{-2} at discharge rates of 1.0 and 2.0 mA. Notably, no deterioration in the cell performance was observed under the bent state (bending diameter = 0.95 ~ 3.81 cm).

Printed thin-film batteries are expected to be combined with printed electronics using the same printing techniques, eventually leading to a monolithically integrated electronics/power source device. A challenge in this integrated device is the low cell potential arising from the alkaline battery (~1.5 V), which is not sufficient to fulfil the requirement of high working potential for printed electronics (e.g., thin-film transistors (10 ~ 30 V)) [55, 56]. To resolve this

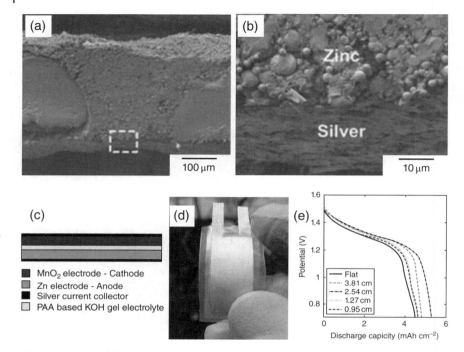

Figure 5.8 Printed, flexible Zn/MnO$_2$ batteries based on Nylon mesh-containing electrodes. (a) SEM image (cross-section) of the mesh-embedded Zn electrode and (b) high-resolution SEM image of the Zn-silver interface. (c) Schematic diagram of the assembled Zn/MnO$_2$ alkaline battery with in-series configuration. (d) Photograph of the flexible Zn/MnO$_2$ battery laminated inside a polyethylene pouch. (e) Discharge profiles of the flexible Zn/MnO$_2$ battery at a constant current of 1 mA under the bent state (bending diameter = 0.95–3.81 cm). Reproduced with permission [54].

problem, a series-connection of alkaline cells was suggested (Figure 5.9) [32]. The stencil-printed unit cells were connected in series using silver interconnects, providing a total potential of ~ 14 V, which is high enough to power an inkjet-printed complementary 5-stage ring oscillator.

The aforementioned Zn/MnO$_2$ batteries still failed to achieve reliable charge/ discharge cyclability. Wright *et al.* reported a rechargeable Zn/MnO$_2$ microbattery by use of a direct-write dispenser-printing technique [49]. However, shape change in the zinc electrode, dendrite formation, and dissolution of reaction products into electrolytes remained formidable challenges in securing meaningful electrochemical rechargeability. To address this issue, they used ionic liquids as an alternative electrolyte [57]. The resulting battery showed an areal capacity of 0.98 mAh cm^{-2} over 70 cycles, although the mechanism for such improved cyclability was not fully elucidated.

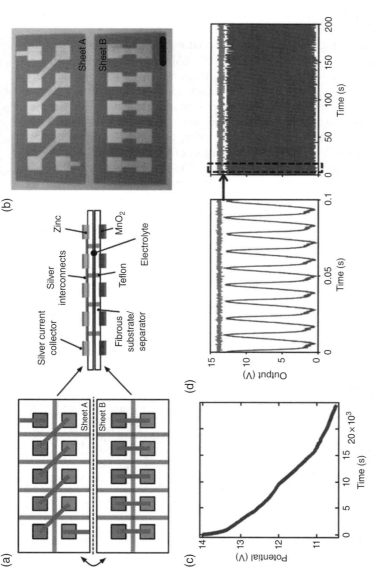

Figure 5.9 (a) Schematic illustration of the Zn-MnO₂ battery with 10 series-connected cells, which was printed on a fibrous substrate. The printed silver electrodes served as the current collectors and interconnects. Amorphous fluoropolymer solution printed inbetween the Zn and MnO₂ electrodes segregated the electrolyte in individual cells. The resultant printed membrane was cut into two sheets (sheets A and B). Each electrode in sheets A and B was soaked in KOH/ZnO electrolyte and stacked together to complete the battery. (b) Photograph of sheets A and B after printing Zn, MnO₂ electrodes, silver current collectors and interconnects (scale bar = 2 cm). (c) Discharge profile of the battery under the load of a 100 kΩ resistor. (d) Output from the printed 5-stage ring oscillator when powered with a 14 V printed battery (blue line: oscillator output, red line: battery output). Reproduced with permission [32].

5.3.2 Supercapacitors

Printed supercapacitors were fabricated through a variety of printing techniques such as screen printing [58], inkjet printing [2, 59, 60], and spraying [61, 62]. Cui *et al.* presented printed thin-film supercapacitors based on CNT networks (serving as both electrodes and current collectors) and printed gel polymer electrolyte [61]. The printed supercapacitors showed high energy (6 Wh kg^{-1}) and power densities (70 kW kg^{-1}) which were comparable to the results of other CNT-based supercapacitors.

Gogotsi *et al.* demonstrated textile supercapacitors based on knitted carbon fibers and activated carbon (AC) ink (Figure 5.10) [63]. The textile supercapacitors showed capacitances as high as 0.5 F cm^{-2} per device at 10 mV s^{-1}, which were comparable with those of conventional AC film electrodes. Also, the textile supercapacitor exhibited excellent electrochemical performance when bent and stretched.

Lee *et al.* demonstrated solid-state flexible supercapacitors which were fabricated directly on conventional A4 paper using a household desktop inkjet printer (Figure 5.11) [2]. The new supercapacitors looked like typical inkjet-printed letters or patterns commonly found in office documents. To attain high-resolution print images on A4 paper, a cellulose nanofibril (CNF) nanomat was inkjet-printed on A4 paper in advance as a primer layer. The Ag nanowires were introduced onto the single-walled carbon nanotubes (SWNT)/AC electrodes in order to improve electrical conductivity of the resulting inkjet-printed

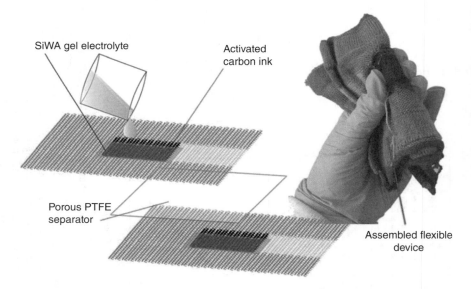

Figure 5.10 Schematic representation of textile supercapacitors based on knitted carbon fibers and activated carbon inks. Reproduced with permission [63]. (*See insert for color representation of the figure.*)

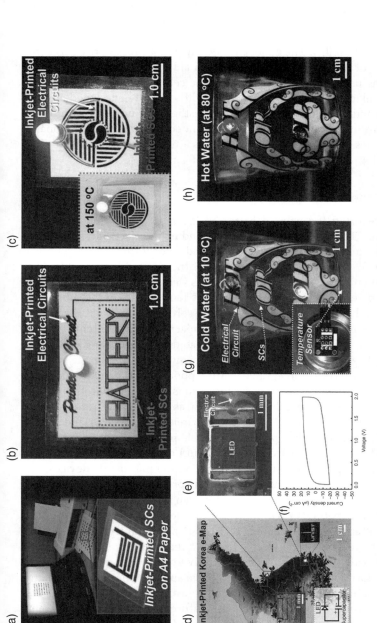

Figure 5.11 All-inkjet-printed, solid-state flexible supercapacitors (SCs) on A4 paper. (a) Photograph of the inkjet-printed SCs. (b) Photograph of the inkjet-printed, letter-shaped SC that was seamlessly connected with the inkjet-printed electrical circuits and an LED lamp. (c) Photograph of the inkjet-printed, traditional Korean "Taegeuk" symbol-like SC that was seamlessly connected with the inkjet-printed electrical circuits and an LED lamp. (d) Photograph of the inkjet-printed Korea map, wherein the inkjet-printed SCs were seamlessly connected to LED lamps via the inkjet-printed electrical circuits. (e) SEM image of the LED lamp connected to the inkjet-printed electric circuits. (f) CV profile (scan rate = 1.0 mV s^{-1}) of the inkjet-printed SC in the map. (g) Photograph depicting the operation of the blue LED lamp in the smart cup (with cold water (~10°C)), wherein the inset is a photograph of a temperature sensor assembled with an Arduino board. (h) Photograph depicting the operation of the red LED lamp in the smart cup (with hot water (~80°C)). Reproduced with permission [2].

electrodes, which was followed by SWNT-assisted photonic interwelding of the Ag nanowire networks. The ([BMIM][BF$_4$]/ETPTA) solid-state electrolyte was applied to the inkjet-printed electrodes via the same inkjet printing, followed by the UV-crosslinking process for solidification. The inkjet-printed supercapacitors exhibited reliable electrochemical performance over 2,000 cycles as well as good mechanical flexibility. The inkjet-printed supercapacitors were easily connected in series or in parallel without the extra aid of metallic interconnects, thereby enabling user-customized control of the cell voltage and capacitance. More notably, inkjet-printed supercapacitors with computer-designed artistic patterns and letters were fabricated and also aesthetically unitized with other inkjet-printed art images and smart glass cups, underscoring their exceptionally versatile aesthetics and potential applicability as a new class of object-tailored power sources for use in forthcoming Internet of Things objects.

5.3.3 Li-ion Batteries

Ajayan *et al.* reported the painted battery using a spray-printing technique (Figure 5.12) [44]. The battery components, including the electrodes, separator membranes, and current collectors, were prepared in the form of paint solutions and then spray-coated sequentially to construct the painted battery [44]. The paints of the aforementioned battery components were sprayed through shadow masks with predetermined geometries, while the temperature of the substrates was controlled in the range 90 ~ 120 °C to remove solvents in the painted components. The painted battery provided an areal capacity of > 1 mAh cm^{-2} with a working voltage of ~ 2.3 V. The battery was fabricated with a large area of 25 cm^2, demonstrating the scalability of the adopted spraying process. Also, the painted battery was expected to be integrated with a photovoltaic panel to develop energy conversion/storage hybrid devices for various outdoor applications.

The proliferation of microscale devices, such as micro-electro-mechanical systems (MEMS) [64], biomedical sensors [65], wireless sensors [66], and actuators [67], prompts the relentless pursuit of flexible power sources with various form factors. The 3D battery based on micro- and nano-structured electrodes could be an appealing power source for operating the microscale devices. To date, however, the 3D architectures have been produced in planar and 3D motifs using complicated/multi-step processes such as lithography [68–70] and colloidal templating methods [71]. To overcome such process complexity, 3D-printed microbatteries were presented [12, 51]. 3D-printed microbatteries were prepared with high-aspect-ratio electrode arrays that were interdigitated on a sub-millimeter scale (Figure 5.13). The relatively low power density, however, has remained a formidable challenge for practical applications; it is ascribed to the large interdigital distance that is needed to avoid short-circuit problems between the electrodes. Moreover, high temperature post treatment was essentially required during fabrication of the 3D-printed electrodes, thus imposing critical restrictions on widening the material design and formulation.

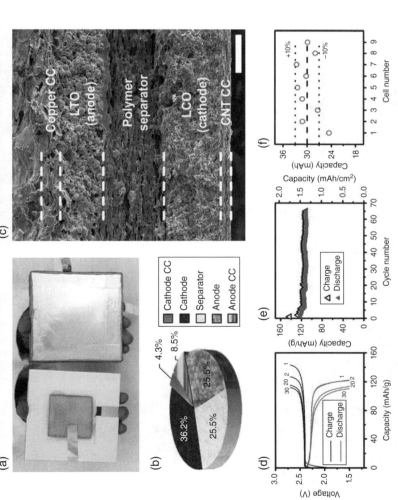

Figure 5.12 Spray-printed lithium-ion batteries. (a) (Left) Glazed ceramic tile with spray-printed lithium-ion cell (area = 5 × 5 cm², capacity = 30 mAh). (Right) Similar cell packaged with laminated PE-Al-PET sheets after electrolyte addition and heat sealing. (b) Mass distribution of components in the spray-printed lithium-ion battery. (c) Cross-sectional SEM image of the spray-printed full cell showing its multilayered structure, with interfaces between successive layers indicated by dashed lines for clarity (Scale bar is 100 μm). (d) Charge/discharge curves for 1st and 30th cycles for the spray-printed full cell (LiCoO₂/Li₄Ti₅O₁₂) cycled at a rate of C/8 between 2.7 and 1.5V. (e) Specific capacity vs. cycle numbers for the spray-printed full cell. (f) Capacities of 8 out of 9 cells fall within 10% of the targeted capacity of 30 mAh, suggesting good process control over a complex device even with manual spray printing. Reproduced with permission [44].

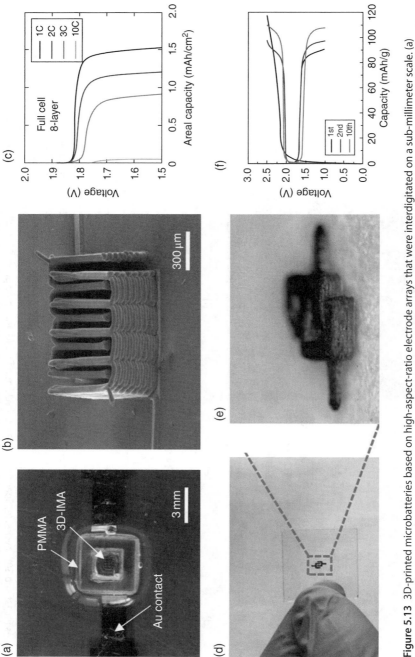

Figure 5.13 3D-printed microbatteries based on high-aspect-ratio electrode arrays that were interdigitated on a sub-millimeter scale. (a) Photograph of the 3D-printed battery after packaging. (b) SEM image of the 3D-printed, 16-layer interdigitated electrode. (c) Discharge profiles of the 3D-printed microbattery as a function of areal capacity. (d, e) Digital images of a miniaturized version of the 3D-printed graphene electrodes. (f) Charge/discharge profiles of the 3D-printed full cell. Reproduced with permission [12, 51].

The abovementioned previous studies showed meaningful achievements in cell design and architecture. However, most used pre-designed masks, supplementary spatial alignments, and liquid electrolytes as an ionic medium. In particular, the use of liquid electrolytes gives rise to problems such as safety failure, an additional electrolyte injection step, and packaging limitation. To address this long-standing issue related to liquid electrolyte, Lee *et al.* reported printed, solid-state rechargeable lithium-ion batteries (Figure 5.14) [1]. The UV-cured

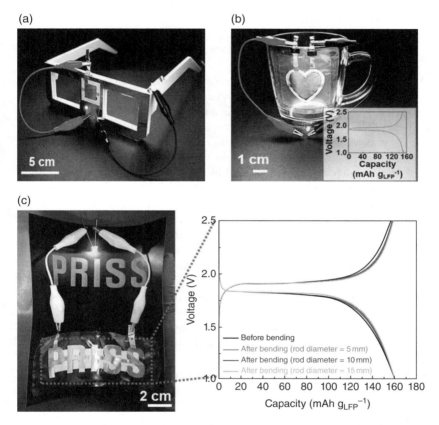

Figure 5.14 Printed, solid-state lithium-ion batteries. (a) Photograph showing direct fabrication and operation of the printed lithium-ion cell on paper-made eyeglasses. (b) Photograph showing direct fabrication of the printed lithium-ion cell on a transparent glass cup with curvilinear surface. The printed lithium-ion cell, having being mounted on the round glass cup, delivered normal charge/discharge behavior (at charge/discharge current density of 0.05 C/0.05 C under a voltage range of 1.0 – 2.5 V). (c) Photograph of "PRISS" letters-shaped, printed lithium-ion cell (left side) and its charge/discharge profiles at charge/discharge current density of 0.05 C/0.05 C under voltage range of 1.0 – 2.5 V (right side), which were measured having being completely wound along rods with different diameters (=5, 10, 15 mm). Reproduced with permission [1]. (*See insert for color representation of the figure.*)

solid-state composite electrolyte was incorporated in the printed electrodes, in addition to acting as a printed separator membrane and electrolyte. As a consequence, time-/energy-consuming solvent-drying processes and electrolyte injection steps were eliminated in the cell fabrication. All of the cell components were sequentially printed, followed by UV irradiation for solidification, resulting in the fabrication of all-solid-state printed batteries. Elaborate combination of the printing technology with the solid-state electrolyte enabled the seamless integration of the batteries with various objects such as paper-made eyeglasses and glass cups.

5.3.4 Other Systems

Zn/Ag batteries have attracted considerable attention due to their high energy density ($200 \, Wh \, kg^{-1}$ and $750 \, Wh \, L^{-1}$) and air stability [36]. However, the high material cost of silver poses a barrier to broadening the application field of Zn/Ag batteries. The printing technology can be suggested as an efficient way to reduce the mass consumption of silver, thus boosting research into Zn/Ag batteries as a promising power source system.

Wang *et al.* reported a rechargeable, skin-worn Ag-Zn tattoo battery, which consisted of screen-printed electrodes, temporary tattoo paper, alkaline gel electrolytes and a PDMS cover (Figure 5.15) [52]. The tattoo battery showed an areal capacity in the range 1.3–$2.1 \, mAh \, cm^{-2}$ over 13 cycles. The open-circuit voltage of $1.5 \, V$ was retained over 5 days upon repeated stretching and bending strain cycles. The lateral arrangement of the cathode and anode allowed the integration of several cells in series or in parallel, resulting in facile tuning of capacity and voltage as initially designed. The application of the tattoo battery to human skin as a shape-conformable power source was successfully achieved.

Subramanian *et al.* demonstrated the development of a photopolymerizable alkaline separator based on poly(acrylic acid) (Figure 5.16) [37]. This PAA separator showed good mechanical stability and ionic conductivity of $0.4 \, S \, cm^{-1}$. Using the PAA separator as a core component, they presented a stencil-printed, silver oxide (Ag_2O)-Zn battery with areal capacities of $5.4 \, mAh \, cm^{-2}$ and volumetric capacities of $7.1 \, mAh \, cm^{-3}$. The batteries showed no loss in electrochemical performance upon flexing at a bend radius of $1.0 \, cm$, indicating their potential use in flexible electronics.

5.4 Summary and Prospects

In summary, we have reviewed the design principle and recent research progress in the field of printed batteries, with a particular focus on slurry/ink chemistry, printing techniques, and aesthetic versatility.

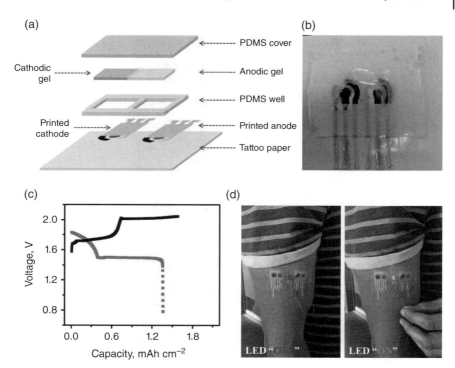

Figure 5.15 Rechargeable, skin-worn Ag-Zn tattoo battery. (a) Schematic illustration depicting the fabrication steps of the Zn/Ag tattoo cell. (b) Photograph of the Zn/Ag tattoo cell on a temporary transfer tattoo support. (c) Charge/discharge profile of the Zn/Ag tattoo cell. (d) Application of the Zn/Ag tattoo cell onto the skin. Reproduced with permission [52].

A first step in the development of printed batteries is the preparation of battery component slurries/inks, which should be finely tuned to ensure process compatibility with printing technologies. In printed batteries, the electrodes should be electrochemically active, indicating that ionic channels as well as electronic networks must be simultaneously secured. However, most of the previously reported electrode slurries/inks have only had electrically conductive components, such as carbon additives, without ionically conductive ones. These electrodes should additionally uptake liquid electrolytes for electrochemical activation after being assembled with separator membranes/packaging substances. This poses a formidable challenge for the development of printed batteries with various form factors. A new approach to fabricating electrode slurries with mixed electron/ion conduction was demonstrated. The electrode slurries were composed of electrode active powders, carbon-conductive additives, and ionically conductive polymer matrix, thereby eliminating the liquid-electrolyte injection step and also traditional processing solvents that demand the time-/energy-consuming drying step.

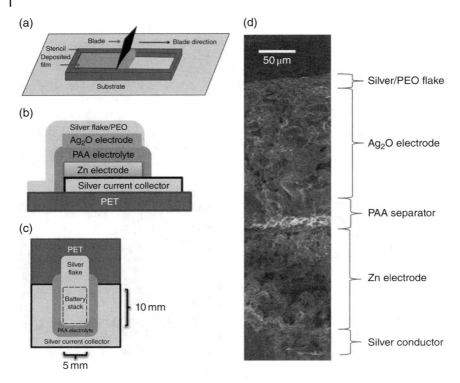

Figure 5.16 Stencil-printed, Ag2O-Zn battery. (a) Schematic representation depicting the stencil printing of the electrode slurry. (b) Cross-sectional illustration of the stencil-printed Ag2O-Zn battery. (c) Top-down view of the stencil-printed Ag2O-Zn battery stack. (d) SEM image of the stencil-printed Ag2O-Zn battery stack. Reproduced with permission [37].

Conventional batteries have (cylindrical-/rectangular-shaped) forms which are generally fabricated using a winding or stacking process. Such stereotypical cell assembly has meant that the power sources lack a variety of form factors and design flexibility. This unwanted limitation is believed to mainly arise from the use of sheet-type separator membranes and liquid electrolytes. In most of the printed batteries investigated so far, separator membranes and electrolytes have not been a center of attention as compared to electrodes. To reach the ultimate goal of so-called "all-printed-batteries", printed separator membranes/solid-state electrolytes should be also developed. Through the combination of printed separator membranes/solid-state electrolytes with printed electrodes, the resultant printed batteries enabled unprecedented advances in form factors and aesthetic versatility. Representative examples include letter-shaped batteries, batteries with in-series or in-parallel configurations, and batteries monolithically integrated into artistic electronic patterns and art images.

Printed batteries have garnered great attention as an innovative technology that breaks common beliefs around power sources from the design perspective. Future research into printed batteries will be directed to the design of new battery component slurries/inks based on chemical/rheological understanding and the adoption of newly emerging printing technologies such as high-fidelity inkjet, electrohydrodynamic, and 3D printing. These will eventually enable a new class of object-tailored power sources with exceptional design versatility for flexible/wearable electronics and IoT.

Acknowledgements

This work was supported by the Basic Science Research Program (2015R1A2A1A01003474) and Wearable Platform Materials Technology Center (2016R1A5A1009926) through the National Research Foundation of Korea (NRF) funded by the Ministry of Science, ICT and future Planning.

References

1 Kim, S.H., Choi, K.H., Cho, S.J., Choi, S., Park, S., Lee, S.Y. (2015) *Nano Lett.* **15**, 5168.

2 Choi, K.-H., Yoo, J., Lee, C.K., Lee, S.-Y. (2016) *Energy Environ. Sci.* **9**, 2812.

3 Gaikwad, A.M., Arias, A.C., Steingart, D.A. (2015) *Energy Technol.* **3**, 305.

4 Sousa, R.E., Costa, C.M., Lanceros-Mendez, S. (2015) *ChemSusChem* **8**, 3539.

5 Tian, D., Song, Y., Jiang, L. (2013) *Chem. Soc. Rev.* **42**, 5184.

6 Hu, L., Cui, Y. (2012) *Energy Environ. Sci.* **5**, 6423.

7 Lawes, S., Riese, A., Sun, Q., Cheng, N., Sun, X. (2015) *Carbon* **92**, 150.

8 Aleeva, Y., Pignataro, B. (2014) *J. Mater. Chem. C* **2**, 6436.

9 Bonaccorso, F., Bartolotta, A., Coleman, J.N., Backes, C. (2016) *Adv. Mater.* **28**, 6136.

10 Farahani, R.D., Dube, M., Therriault, D. (2016) *Adv. Mater.* **28**, 5794.

11 Pech, D., Brunet, M., Taberna, P.-L., Simon, P., Fabre, N., Mesnilgrente, F. *et al.* (2010) *J. Power Sources* **195**, 1266.

12 Sun, K., Wei, T.S., Ahn, B.Y., Seo, J.Y., Dillon, S.J., Lewis, J.A. (2013) *Adv. Mater.* **25**, 4539.

13 Delannoy, P.E., Riou, B., Brousse, T., Le Bideau, J., Guyomard, D., Lestriez, B. (2015) *J. Power Sources* **287**, 261.

14 Milroy, C., Manthiram, A. (2016) *Chem. Commun.* **52**, 4282.

15 Jung, C.Y., Zhao, T.S., An, L., Zeng, L., Wei, Z.H. (2015) *J. Power Sources* **297**, 174.

16 El Baradai, O., Beneventi, D., Alloin, F., Bongiovanni, R., Bruas-Reverdy, N., Bultel, Y. *et al.* (2016) *J. Mater. Sci. Technol.* **32**, 566.

17 Gu, Y., Wu, A., Sohn, H., Nicoletti, C., Iqbal, Z., Federici, J.F. (2015) *J. Manuf. Processes* **20**, 198.
18 Sousa, R.E., Oliveira, J., Gören, A., Miranda, D., Silva, M.M., Hilliou, L. *et al.* (2016) *Electrochim. Acta* **196**, 92.
19 Prosini, P.P., Mancini, R., Petrucci, L., Contini, V., Villano, P. (2001) *Solid State Ionics* **144**, 185.
20 Lee, S.-T., Jeon, S.-W., Yoo, B.-J., Choi, S.-D., Kim, H.-J., Lee, S.-M. (2006) *J. Power Sources* **155**, 375.
21 Zhao, Y., Zhou, Q., Liu, L., Xu, J., Yan, M., Jiang, Z. (2006) *Electrochim. Acta* **51**, 2639.
22 Park, M.S., Hyun, S.-H., Nam, S.-C. (2006) *J. Power Sources* **159**, 1416.
23 Park, M.-S., Hyun, S.-H., Nam, S.-C. (2007) *Electrochim. Acta* **52**, 7895.
24 Huang, J., Yang, J., Li, W., Cai, W., Jiang, Z. (2008) *Thin Solid Films* **516**, 3314.
25 Wei, D., Andrew, P., Yang, H., Jiang, Y., Li, F., Shan, C. *et al.* (2011) *J. Mater. Chem.* **21**, 9762.
26 Izumi, A., Sanada, M., Furuichi, K., Teraki, K., Matsuda, T., Hiramatsu, K. *et al.* (2012) *Electrochim. Acta* **79**, 218.
27 Kang, K.-Y., Lee, Y.G., Shin, D.O., Kim, J.C., Kim, K.M. (2014) *Electrochim. Acta* **138**, 294.
28 Izumi, A., Sanada, M., Furuichi, K., Teraki, K., Matsuda, T., Hiramatsu, K. *et al.* (2014) *J. Power Sources* **256**, 244.
29 Tam, W.G., Wainright, J.S. (2007) *J. Power Sources* **165**, 481.
30 Gaikwad, A.M., Khau, B.V., Davies, G., Hertzberg, B., Steingart, D.A., Arias, A.C. (2014) *Adv. Energy Mater.* 1401389.
31 Hu, L., Wu, H., La Mantia, F., Yang, Y., Cui, Y. (2010) *ACS Nano* **4**, 5843.
32 Gaikwad, A.M., Steingart, D.A., Nga Ng, T., Schwartz, D.E., Whiting, G.L. (2013) *Appl. Phys. Lett.* **102**, 233302.
33 Ghiurcan, G.A., Liu, C.-C., Webber, A., Feddrix, F.H. (2003) *J. Electrochem. Soc.* **150**, A922.
34 Winslow, R., Wu, C.H., Wang, Z., Kim, B., Keif, M., Evans, M., Wright, P.K. (2013) *JPCS* **476**, 012085.
35 Wang, Z., Winslow, R., Madan, D., Wright, P.K., Evans, J.W., Keif, M. *et al.* (2014) *J. Power Sources* **268**, 246.
36 Braam, K.T., Volkman, S.K., Subramanian, V. (2012) *J. Power Sources* **199**, 367.
37 Braam, K., Subramanian, V. (2015) *Adv. Mater.* **27**, 689.
38 Ho, C.C., Murata, K., Steingart, D.A., Evans, J.W., Wright, P.K. (2009) *J. Micromech. Microeng.* **19**, 094013.
39 Fu, J., Lee, D.U., Hassan, F.M., Yang, L., Bai, Z., Park, M.G. *et al.* (2015) *Adv. Mater.* **27**, 5617.
40 Hilder, M., Winther-Jensen, B., Clark, N.B. (2009) *J. Power Sources* **194**, 1135.
41 Pignon, F., Magnin, A., Piau, J.-M. (1998) *J. Rheol.* **42**, 1349.
42 Wallevik, J.E. (2009) *Cem. Concr. Res.* **39**, 14.
43 Hu, L., Choi, J.W., Yang, Y., Jeong, S., La Mantia, F., Cui, L.F. *et al.* (2009) *Proc. Natl. Acad. Sci. U.S.A.* **106**, 21490.

44 Singh, N., Galande, C., Miranda, A., Mathkar, A., Gao, W., Reddy, A.L. *et al.* (2012) *Sci. Rep.* **2**, 481.

45 Kuang, M., Wang, L., Song, Y. (2014) *Adv. Mater.* **26**, 6950.

46 Lessing, J., Glavan, A.C., Walker, S.B., Keplinger, C., Lewis, J.A., Whitesides, G.M. (2014) *Adv. Mater.* **26**, 4677.

47 Kil, E.H., Choi, K.H., Ha, H.J., Xu, S., Rogers, J.A., Kim, M.R. *et al.* (2013) *Adv. Mater.* **25**, 1395.

48 Kim, S.H., Choi, K.H., Cho, S.J., Kil, E.H., Lee, S.Y. (2013) *J. Mater. Chem. A* **1**, 4949.

49 Ho, C.C., Evans, J.W., Wright, P.K. (2010) *J. Micromech. Microeng.* **20**, 104009.

50 Tehrani, Z., Korochkina, T., Govindarajan, S., Thomas, D.J., O'Mahony, J., Kettle, J. *et al.* (2015) *Org. Electron.* **26**, 386.

51 Fu, K., Wang, Y., Yan, C., Yao, Y., Chen, Y., Dai, J. *et al.* (2016) *Adv. Mater.* **28**, 2587.

52 Berchmans, S., Bandodkar, A.J., Jia, W., Ramírez, J., Meng, Y.S., Wang, J. (2014) *J. Mater. Chem. A* **2**, 15788.

53 Kiebele, A., Gruner, G. (2007) *Appl. Phys. Lett.* **91**, 144104.

54 Gaikwad, A.M., Whiting, G.L., Steingart, D.A., Arias, A.C. (2011) *Adv. Mater.* **23**, 3251.

55 Ng, T.N., Schwartz, D.E., Lavery, L.L., Whiting, G.L., Russo, B., Krusor, B. *et al.* (2012) *Sci. Rep.* **2**, 585.

56 Ng, T.N., Sambandan, S., Jujan, R., Arias, A.C., Newman, C., Yan, H., Facchetti, A. (2009) *Appl. Phys. Lett.* **94**, 233307.

57 Kumar, G.G., Sampath, S. (2003) *J. Electrochem. Soc.* **150**, A608.

58 Xu, Y., Schwab, M.G., Strudwick, A.J., Hennig, I., Feng, X., Wu, Z. *et al.* (2013) *Adv. Energy Mater.* **3**, 1035.

59 Wang, S., Liu, N., Tao, J., Yang, C., Liu, W., Shi, Y. *et al.* (2015) *J. Mater. Chem. A* **3**, 2407.

60 Ujjain, S.K., Bhatia, R., Ahuja, P., Attri, P. (2015) *PLoS One* **10**, e0131475.

61 Kaempgen, M., Chan, C.K., Ma, J., Cui, Y., Gruner, G. (2009) *Nano Lett.* **9**, 1872.

62 Huang, C., Young, N.P., Grant, P.S. (2014) *J. Mater. Chem. A* **2**, 11022.

63 Jost, K., Stenger, D., Perez, C.R., McDonough, J.K., Lian, K., Gogotsi, Y. *et al.* (2013) *Energy Environ. Sci.* **6**, 2698.

64 Spearing, S.M. (2000) *Acta Mater.* **48**, 179.

65 Zhang, C., Xu, J., Ma, W., Zheng, W. (2006) *Biotechnol. Adv.* **24**, 243.

66 Fowler, J.D., Allen, M.J., Tung, V.C., Yang, Y., Kaner, R.B., Weiller, B.H. (2009) *ACS Nano* **3**, 301.

67 Waggoner, P.S., Craighead, H.G. (2007) *Lab Chip* **7**, 1238.

68 Gowda, S.R., Reddy, A.L.M., Zhan, X., Ajayan, A. (2011) *Nano Lett.* **11**, 3329.

69 Baggetto, L., Niessen, R.A.H., Roozeboom, F., Notten, P.H.L. (2008) *Adv. Funct. Mater.* **18**, 1057.

70 Nathan, M., Golodnitsky, D., Yufit, V., Strauss, E., Ripenbein, T., Shechtman, I. *et al.* (2005) *J. Microelectromech Syst.* **14**, 879.

71 Zhang, H., Yu, X., Braun, P.V. (2011) *Nat. Nanotechnol.* **6**, 277.

6

Applications of Printed Batteries

Abhinav M. Gaikwad, Aminy E. Ostfeld and Ana Claudia Arias

Electrical Engineering and Computer Sciences Department, University of California, Berkeley, USA

The energy needs of the rapidly evolving consumer electronics industry have consistently pushed the boundaries of battery technology. Until the 1980s, products such as remote controlled toys, handheld radios, torchlights, and flash cameras were the primary consumers of portable batteries. Zinc-Manganese Dioxide ($Zn-MnO_2$)-based alkaline batteries dominated this market for many decades due to their low cost, high energy density, and low toxicity. In the mid-1980s, portable music players and video recorders were introduced in the market, which created a demand for high energy density rechargeable batteries. Rechargeable Ni-Cd batteries emerged during this period, but their adoption was limited due to low energy density, high self-discharge, and toxicity of the cadmium electrode. In 1991, after years of research and development, SONY introduced the world's first rechargeable lithium ion (Li Ion) battery. The batteries were designed to power SONY's highly profitable camcorder line of products. SONY solved the problem of dendrite formation in the anode by replacing the lithium metal electrode with an insertion-based graphite electrode. The cost of the early generations of Li Ion batteries was above 3000$/kWh, which made them suitable only for high-end products. Nickel-Metal Hydride (Ni-MH) batteries were introduced in the mid-1990s and they were adopted in many electronic products. In the 2000s the mobile phone and laptop market grew tremendously. Electronics companies shifted from rechargeable Ni-MH batteries to Li Ion batteries due to their high energy density and cycle life. By the 2010s, the cost of Li Ion batteries dropped below 250$/kWh, which made them economical for all-electric and hybrid vehicles. Tesla motors were the early pioneers in the development of Li Ion battery-powered all-electric

Printed Batteries: Materials, Technologies and Applications, First Edition.
Edited by Senentxu Lanceros-Méndez and Carlos Miguel Costa.
© 2018 John Wiley & Sons Ltd. Published 2018 by John Wiley & Sons Ltd.

vehicles. By 2013, all the major car manufacturers around the world announced projects on battery-powered electric vehicles. The brief history of portable batteries shows that advances in battery technology have always been driven by the development of new products. Over the years the primary consumers of batteries have shifted from toys to cameras, mobile phones, and laptops, and, more recently, to electric vehicles.

In recent years, advances in Microelectromechanical Systems (MEMS) fabrication, low-power microprocessors, big-data analytics and solution-based device processing opened up an array of new products, such as activity trackers, smart clothing, printed electronics and e-skins. These products are part of a large eco-system of devices commonly referred to as 'Internet of Things' (IOT). In this eco-system, a swarm of low-cost, connected devices will be distributed around us. These smart devices will sense, communicate with each other, take action, and provide information to the user. The battery requirements for these products are fundamentally different from those of a laptop or an electric vehicle [1]. In addition to typical characteristics such as low cost and high energy density, the requirements for IOT applications often include characteristics such as small footprint (<1 cm^2), flexibility, unique form factor and even stretchability for certain applications. To satisfy these unique sets of requirements, researchers developed additive printing technologies for fabricating batteries [2–7]. Batteries fabricated with the help of additive printing technologies are collectively called 'printed batteries'. We have seen tremendous progress in this field since the early reports in the mid-2000s. In recent years, most academic battery labs around the world have published papers on novel printing technologies [8, 9], ink formulations [10], and electrode designs [11, 12]. There are a number of startup companies that are commercializing printed batteries. Even the large battery manufacturers such as LG Chem, Samsung, Panasonic, NEC, and Sanyo have demonstrated printed flexible batteries in numerous consumer electronics expos over the last couple of years. The interest from academia and industry demonstrates that there is going be a large demand for printed flexible batteries in the near future.

The term 'printed battery' is used to describe a battery in which most of the components of the battery are deposited with an additive printing technology. Printed batteries are available in a variety of designs, capacities, shapes and chemistries. Table 6.1 gives an overview of the key differences between a conventional and a printed battery. Printed batteries are encapsulated within compliant materials such as plastic, elastomer or metal-laminated plastic, whereas conventional batteries are encapsulated in rigid metal casings. The capacities of printed batteries are much lower in comparison to conventional batteries. An Li Ion battery inside a mobile phone has a capacity in the range of 3000 mAh whereas a printed battery with a similar footprint would have a capacity in the range 50 to 100 mAh. The volumetric energy densities of printed batteries are much lower in comparison to conventional batteries. In many designs

Table 6.1 Differences between printed and conventional batteries.

	Printed battery	Conventional battery
Encapsulation	Plastic, elastomer	Metal casing, metal-laminated pouches
Capacity	10 to 100 mAh	500 to 3000 mAh
Energy density	50 to 300 Wh/L	300 to 700 Wh/L
Flexibility	Bendable up to 3 to 5 cm radius	Not flexible

the encapsulation layer is as thick as the thickness of the active components, which reduces the volume occupied by the active material. Most printed batteries are flexible to some extent due to their thin form factor and flexible encapsulation material. The capacity retention after flexing depends on the design of the battery.

In this chapter, we will focus on the applications of printed batteries. We would like to point out to readers that the field of printed batteries is still in its early stages of development. Applications discussed in this chapter are based on products that are mostly likely to benefit from the characteristics of printed batteries. We divide this chapter into five sections. In the first section, we will discuss the applications of microbatteries. These batteries have a footprint in the mm^2 range and they are useful for powering devices with a very small footprint. In the second section, we will discuss the applications of thin printed primary batteries. These batteries are thin, cheap, easy to fabricate, and useful for powering disposable types of electronics. In the third section, we will discuss the applications of high-performance rechargeable flexible batteries. These batteries have electrochemical performance comparable to conventional batteries. These batteries can be embedded within current generations of wearable products and enable new product designs, designs which are not possible with conventional batteries. In the fourth section, we will discuss the use of printing technologies to fabricate structured electrodes. The structured design helps to improve the energy density and rate performance of the current generation of batteries. In the fifth section, we will discuss the power electronics required for charging printed batteries with harvesters.

6.1 Printed Microbatteries

The advances in microelectromechanical systems (MEMS), sensing technologies, and low-power wireless communications have made it possible to design sensing nodes with a footprint of less than 1 cm^2 [13]. A typical autonomous sensing node contains a couple of sensors, memory, microcontroller, wireless

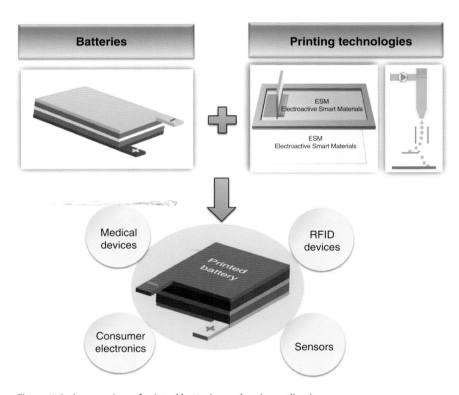

Figure 1.2 An overview of printed batteries and main applications.

Printed Batteries: Materials, Technologies and Applications, First Edition.
Edited by Senentxu Lanceros-Méndez and Carlos Miguel Costa.
© 2018 John Wiley & Sons Ltd. Published 2018 by John Wiley & Sons Ltd.

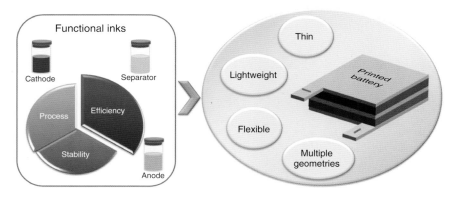

Figure 1.3 An overview of the functional inks and relevant requirements in the area of printed battery research.

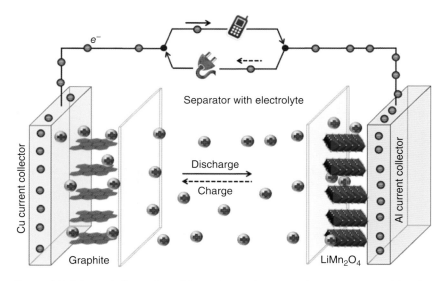

Figure 1.5 Schematic illustration of the main constituents and representation of the charge and discharge modes of a battery.

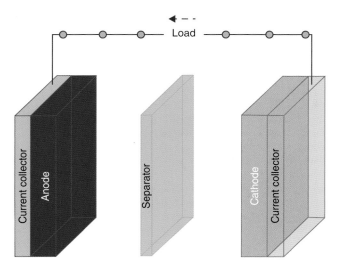

Figure 2.2 Schematic representation of the main components of a battery.

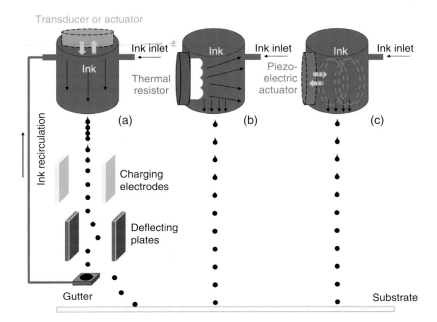

Figure 2.14 Scheme showing the basic principles of (a) CIJ technology, DoD-based (b) TIJ and (c) PIJ technology.

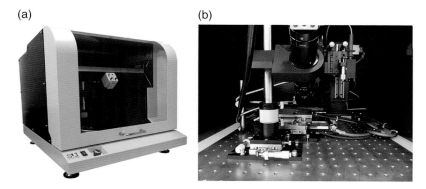

Figure 2.17 SIJ inkjet printer SIJ-S050 from SIJTechnology Inc. (a) full machine, (b) inside view [70, 71].

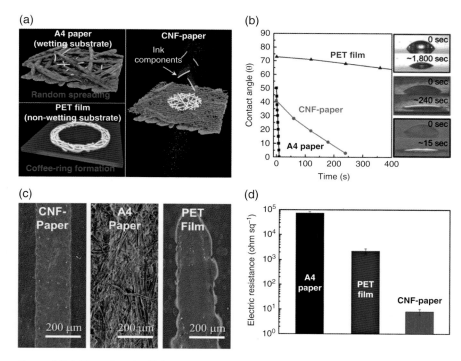

Figure 5.3 Inkjet-printed cellulose nanomat on commercial A4 paper. (a) Effect of substrates on the resolution of the inkjet printing process: the wetting substrate (left upper side, random spreading of ink droplets), the non-wetting substrate (left lower side, coffee-ring formation), and the CNF nanomat on A4 paper (right side, high-resolution print pattern). (b) Variation in the water-contact angles of different substrates with time. (c) SEM images (surface view) of the inkjet-printed ((SWNT/AC) + Ag NWs) electrodes on different substrates. (d) Electric resistances of the inkjet-printed ((SWNT/AC) + Ag NWs) electrodes on different substrates. Reproduced with permission [2].

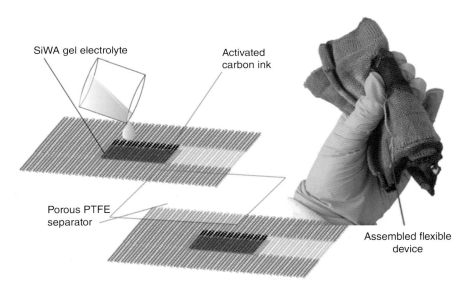

SiWA gel electrolyte

Activated carbon ink

Porous PTFE separator

Assembled flexible device

Figure 5.10 Schematic representation of textile supercapacitors based on knitted carbon fibers and activated carbon inks. Reproduced with permission [63].

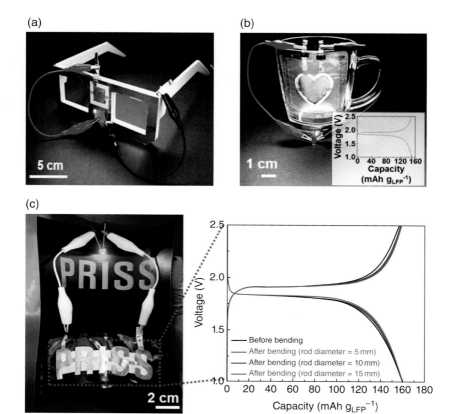

Figure 5.14 Printed, solid-state lithium-ion batteries. (a) Photograph showing direct fabrication and operation of the printed lithium-ion cell on paper-made eyeglasses. (b) Photograph showing direct fabrication of the printed lithium-ion cell on a transparent glass cup with curvilinear surface. The printed lithium-ion cell, having being mounted on the round glass cup, delivered normal charge/discharge behavior (at charge/discharge current density of 0.05 C/0.05 C under a voltage range of 1.0 – 2.5 V). (c) Photograph of "PRISS" letters-shaped, printed lithium-ion cell (left side) and its charge/discharge profiles at charge/discharge current density of 0.05 C/0.05 C under voltage range of 1.0 – 2.5 V (right side), which were measured having being completely wound along rods with different diameters (=5, 10, 15 mm). Reproduced with permission [1].

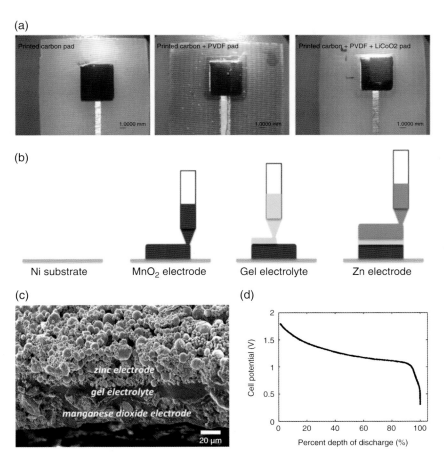

Figure 6.3 (a) Optical images of a dispenser-printed lithium ion polymer battery after the deposition of graphite, PVDF separator and LCO layer [2]. (b) Schematic of the fabrication process of Zn-MnO$_2$ battery. (c) SEM micrograph Zn-MnO$_2$ battery. (d) Discharge characteristics [3]. Copyright (2010), IOP Publishing.

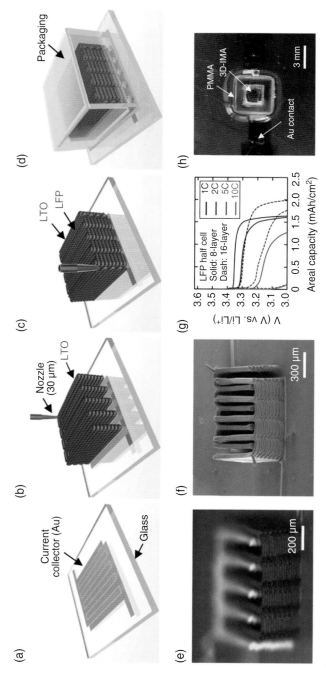

Figure 6.4 (a–d) Schematic of the fabrication process of three-dimensional interdigitated microbattery. The process starts with patterning the gold current collector, followed by depositing the slurry for the anode and cathode, followed by encapsulation and soaking of the electrode with electrolyte. (e, f) SEM micrographs of the microbattery. (g) Rate performance of the microbattery with 8 and 16 layers of electrodes. (h) Photograph of the encapsulated microbattery [16]. Copyright (2013), John Wiley and Sons.

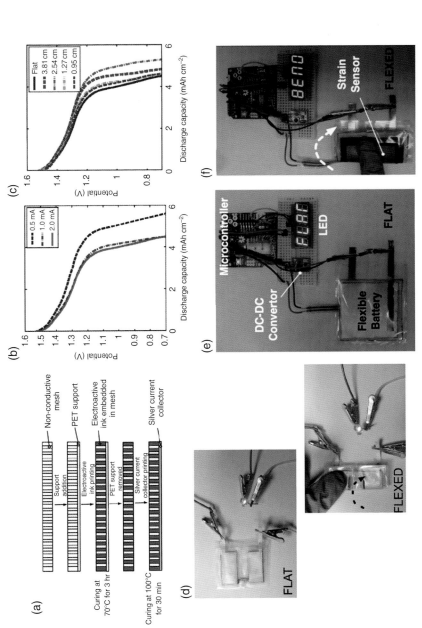

Figure 6.8 (a) Process flow of fabricating a mesh-embedded flexible alkaline battery. Discharge curves at the mesh-battery at various C-rates (b) and discharge curves of the battery when flexed to various bending radii (c). (d) Mesh-battery powering a green-LED under flat and bend conditions [11]. (e) A flexible alkaline battery with reinforced electrode structure powering an interactive display and microcontroller. The display shows FLAT when the battery is flat and the display shows BEND (f) when the battery is bent [31]. Copyright (2013), John Wiley and Sons.

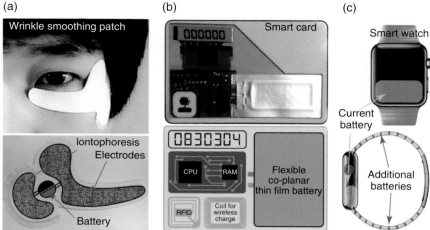

Figure 6.15 Examples of devices powered with a flexible coplanar battery. (a) A cosmetic patch with iontophoresis function powered with a flexible battery, (b) smart bank card to provide additional security for transactions powered with a battery, and (c) smart watch with flexible batteries embedded inside the band of the watch to provide additional capacity [41]. Copyright (2015), American Chemical Society.

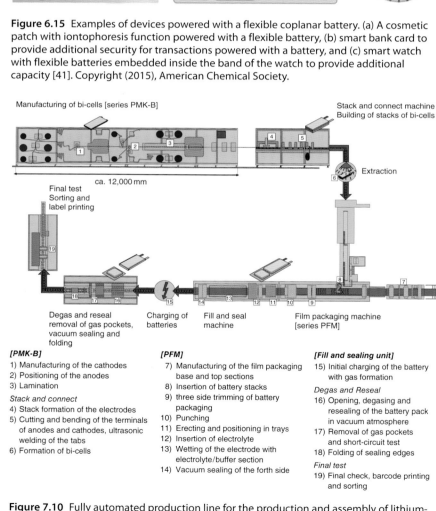

[PMK-B]

1) Manufacturing of the cathodes
2) Positioning of the anodes
3) Lamination

Stack and connect

4) Stack formation of the electrodes
5) Cutting and bending of the terminals of anodes and cathodes, ultrasonic welding of the tabs
6) Formation of bi-cells

[PFM]

7) Manufacturing of the film packaging base and top sections
8) Insertion of battery stacks
9) three side trimming of battery packaging
10) Punching
11) Erecting and positioning in trays
12) Insertion of electrolyte
13) Wetting of the electrode with electrolyte/buffer section
14) Vacuum sealing of the forth side

[Fill and sealing unit]

15) Initial charging of the battery with gas formation

Degas and Reseal

16) Opening, degasing and resealing of the battery pack in vacuum atmosphere
17) Removal of gas pockets and short-circuit test
18) Folding of sealing edges

Final test

19) Final check, barcode printing and sorting

Figure 7.10 Fully automated production line for the production and assembly of lithium-ion batteries. Reproduced with permission of Harro Hoefliger Verpackungsmaschinen GmbH (Allmersbach im Tal, Germany).

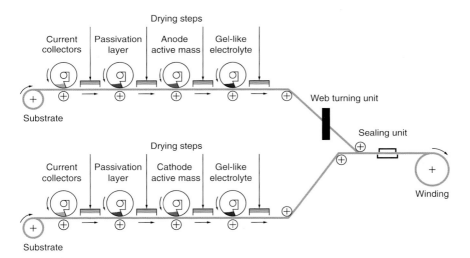

Figure 7.11 Exemplary presentation of two rotary screen printing lines and a sealing unit for the manufacture of zinc-based batteries in stack configuration.

Fully printed coplanar ZnMnO₂ battery

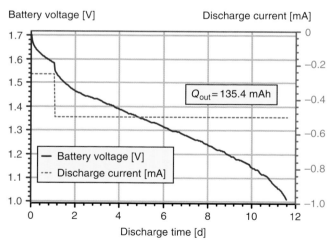

Figure 7.15 Fully printed coplanar zinc-carbon battery (single-cell, ZnCl electrolyte) with an overall thickness of 0.5 mm and the corresponding discharge curve. A capacity of 135.4 mAh was determined during discharge from the open-circuit voltage (OCV) to the cut-off voltage of 1.0 V.

communication, and battery. A large number of these low-cost wireless nodes can be distributed around factories, offices, and homes for sensing and actuation purposes. For example, in a large factory manufacturing a moisture-sensitive product, a network of low-cost sensing nodes can be distributed around the factory to monitor the humidity level. If the humidity level deviates from a set range, the node can send a signal to a central control unit to take action. In another possible application, a number of temperature sensors can be distributed in an office. In a large space it is very common to experience hot and cold zones due to uneven air circulation. By mapping the temperature with a network of temperature sensors, the air circulation can be adjusted to ensure that all the regions in the office are at the same temperature. Such a sensor can also be designed to track the number of occupants in the room. If the room is empty, the heating/cooling can be turned off. Wireless nodes are also finding applications in large chemical and petrochemical plants to track the health of the pumps, pipes, and reactors. Once a problem is detected, the sensor node can notify the maintenance crew to check the equipment and perform necessary repairs. Such sensors can help prevent a potential shutdown of the plant due to equipment failure.

The early generations of wireless sensors were large enough to support a large cylindrical or coin cell. With advances in microfabrication technologies, future generations of wireless nodes are expected to be much smaller in size. For these applications, even the smallest available coin cells in the market are too large in size [5, 14–16]. The need to power micro-sensing nodes drove the early research efforts in developing microbatteries. Bates *et al.* from the Oak Ridge National Laboratory demonstrated the first thin film lithium batteries [17]. In a thin film battery, all the components of the battery, the current collectors, anode, cathode, and separator are deposited at high temperature with a combination of sputtering and vapor deposition processes. Figure 6.1(a) shows a cross-section of the thin film lithium battery. The battery is supported on a rigid support. These batteries are also referred to as solid-state batteries since all the components of the battery are in solid state. Figure 6.1(b) shows the discharge curves of the battery with a footprint of 1 cm^2 at various C-rates. The average areal capacity of the battery is around 0.1 mAh/cm^2 at a voltage of 3.95 V. The solid-state nature of the active layers and the separator limits the speed at which lithium ions can move in the battery. The thicknesses of the electrodes are limited to 3 to 5 micron to reduce mechanical stresses and reduce polarization during charging and discharging. Even with the advantage of high temperature stability, and thin form factor, thin film batteries are not suitable for many applications due to their low capacities and high cost of production.

The disadvantages and cost constraints of solid-state microbatteries were addressed by taking advantage of printing technologies to fabricate microbatteries [2, 5, 15, 16, 18]. Printing processes are well developed, and inexpensive, and only the required amount of material is deposited over the target area.

(a)

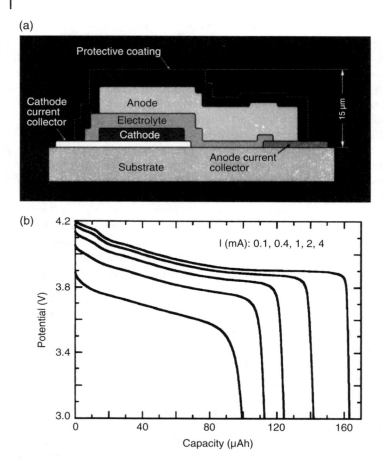

(b)

Figure 6.1 (a) Schematic of cross-section of the thin film battery and (b) discharge curve of the battery at various C-rates [17]. Copyright (2000), Elsevier.

The energy and material consumption required for fabricating printed batteries is much lower in comparison to that of solid-state batteries. Printed microbatteries can be easily customized based on the space available for the battery and the power requirements of the device. The groups of Prof. Paul Wright and James Evans at the University of California, Berkeley, produced initial demonstrations of printed microbatteries [2, 3, 15, 19–21]. A dispenser-based printing technology was adopted to fabricate the printed microbatteries. The fabrication process starts by formulating slurries for the anode, cathode, and separator. The inks are dispensed through needles with diameters ranging from 0.5 to 500 micron. The rheologies of the slurries are adjusted based on the diameter of the needle and the print speed. The ink can be deposited in the

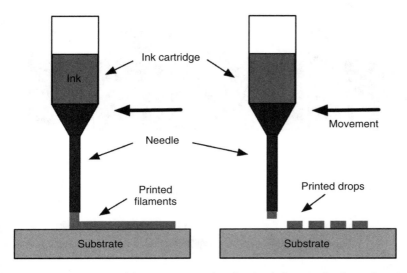

Figure 6.2 Schematic of dispenser-printer head with ink deposited in form of continuous line and drops [1]. Copyright (2015), John Wiley and Sons.

form of drops or line (Figure 6.2). The size of feature depends on the diameter of the needle.

Steingart *et al.* demonstrated a dispenser-printed Li Ion polymer microbattery [2]. In this design, a pre-patterned copper pad on a printed circuit board (PCB) served as a support for the printed layers and as a current collector for the cathode. The fabrication process was carried out by sequentially printing mesocarbon micro beads (MCMB), an ionic liquid with PVDF and lithium cobalt oxide (LCO). Figure 6.3(a) shows optical images of the battery after printing of the graphite, PVDF, and LCO layers. This was one of the earliest demonstrations of dispenser-printed Li Ion microbatteries in which the microbattery was printed directly on a circuit board. This technology can be used to print batteries over dead spaces on a printed circuit board. Christine Ho *et al.* demonstrated a novel chemistry based on $Zn-MnO_2$ and an ionic liquid as the electrolyte [3]. The battery was fabricated with dispenser-printing technology. Thin nickel foils served as the current collector and a support for the MnO_2 electrode. The battery was fabricated by sequentially printing the MnO_2, PVDF-HFP-ionic liquid, and Zn ink on the nickel foil. Figures 6.3(b) and (c) show the schematic of the fabrication process and cross-sectional SEM micrograph of the battery. The thickness of the battery ranged 80–120 microns. The total footprint of the cell was 0.25 cm^2. Figure 6.3(d) shows the discharge characteristics of the battery. The voltage of the charged battery was around 1.8 V and it dropped to 0.4 V at the end of discharge. The capacity of the battery after the activation process was around 1.0 mAh/cm^2. The technology was spun-out

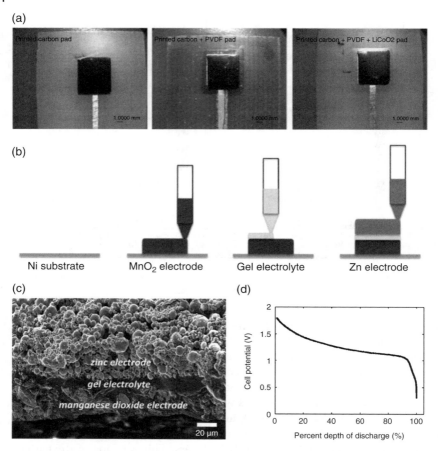

Figure 6.3 (a) Optical images of a dispenser-printed lithium ion polymer battery after the deposition of graphite, PVDF separator and LCO layer [2]. (b) Schematic of the fabrication process of Zn-MnO₂ battery. (c) SEM micrograph Zn-MnO₂ battery. (d) Discharge characteristics [3]. Copyright (2010), IOP Publishing. (*See insert for color representation of the figure.*)

from the University of California at Berkeley. A company called Imprint Energy in the San Francisco Bay Area is commercializing the technology for powering flexible electronics. The company uses a combination of screen and stencil printing to deposit the active layer and polymer separator. It is producing flexible batteries with capacities in the range 1 to 10 mAh/cm². The battery has a capacity retention of above 90% even after thousands of flexing cycles.

A typical wireless sensing node is characterized by long periods of sleep and short active periods during which the device performs sensing action, data processing, and data transfer. The power requirements during the active period are very high, which can lead to significant energy losses due to polarization. Polarization can induce side-reaction within the battery and reduce the lifetime of the battery. In order to reduce polarization in microbatteries and enable high discharge rates, a concept of 3-D architecture was applied to printed microbatteries [8, 13, 16, 19, 22–26]. In a 3-D battery, the rate capability of the battery is improved by increasing the overall surface area of the electrode and by reducing the distance between the anode and cathode. The group of Prof. Jennifer Lewis at the University of Illinois demonstrated the first all-printed three-dimensional interdigitated Li Ion battery [8, 16]. The components of the battery were deposited with a custom dispenser-printer. Multiple layers of the active materials were deposited on top of each other to increase the areal loading of the electrodes. The interdigitated design improved the rate capabilities of the battery. Figure 6.4(a)–(d) shows the schematic of the fabrication process. The process starts by evaporating and patterning gold electrodes on a glass substrate to form the current collectors for the anode and cathode. Thick slurries of lithium iron phosphate and lithium titanate oxide were printed on the gold electrodes with a dispenser-printer. The slurries were formulated to ensure they retained their tubular shape after dispensing from the needle. Multiple layers of the electrodes were printed on top of each other. The battery is encapsulated within a PMMA-based enclosure. Figures 6.4(e) and 6.4(f) show SEM micrograph of the electrodes after the deposition process. Figure 6.4(g) shows the discharge capacities of the batteries with 8 and 16 layers of electrodes at various discharge rates. Figure 6.4(h) shows the optical image of the battery after encapsulation. Such battery can be used as a power source for sensing nodes that required large quantities of power but have a small volume.

There are a number of challenges that need to be addressed before printed microbatteries can be adopted in commercial products. Most printed microbatteries demonstrated in the literature are based on dispenser-printing technology [1]. Dispenser-printing is slow in comparison to traditional large-area printing methods such as slot-die and gravure printing. Printed microbatteries can benefit from the development of new printing technology, which will speed up the fabricating process [1]. Currently the size of the encapsulation for microbatteries is much larger than the size of the battery. The development of a thin printable encapsulation material is necessary to reduce the volume of the battery.

6.2 Printed Primary Batteries

Primary batteries based on zinc chemistry have a distinct advantage over Li Ion batteries due to their low cost, convenience, and environmentally safe materials. Primary batteries were commonly used in electronic products such

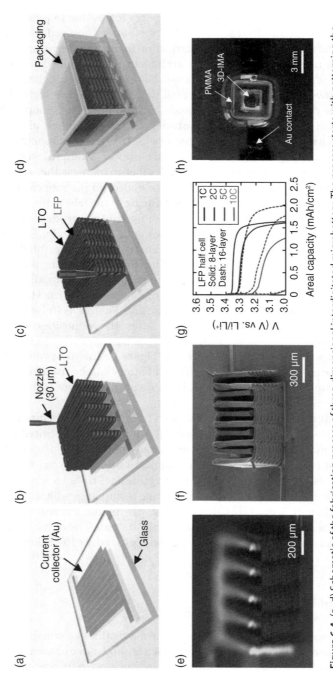

Figure 6.4 (a–d) Schematic of the fabrication process of three-dimensional interdigitated microbattery. The process starts with patterning the gold current collector, followed by depositing the slurry for the anode and cathode, followed by encapsulation and soaking of the electrode with electrolyte. (e, f) SEM micrographs of the microbattery. (g) Rate performance of the microbattery with 8 and 16 layers of electrodes. (h) Photograph of the encapsulated microbattery [16]. Copyright (2013), John Wiley and Sons. (*See insert for color representation of the figure.*)

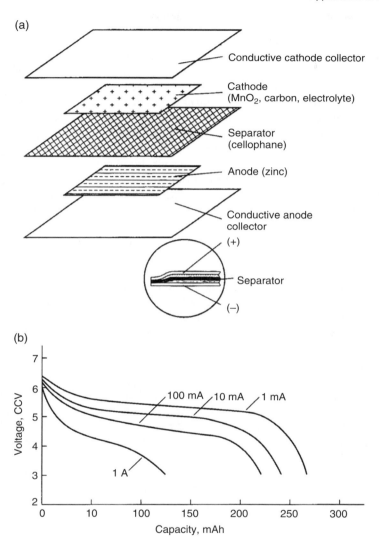

Figure 6.5 (a) Schematic of the P-80 Polaroid battery and (b) discharge curve of the battery pack at various C-rates.

as toys, radios, watches, and cameras before the introduction of Ni-MH and Li Ion batteries. One of the early demonstrations of a printed primary battery was the 'Flat-Pack Zinc-Manganese Dioxide P-80 Battery' developed by Polaroid in the early 1970s for their Instant Camera line of products [27]. Figure 6.5(a) shows the cross-section schematic of the P-80 battery. The major innovation in this product was integration of a flat battery with the film pack. Until recently it was very common for the user to carry an extra set of batteries while operating a camera. The integration of the battery with the film pack meant that the user did not have to worry about carrying extra batteries. The flat battery design was a major engineering achievement. Most primary batteries at the time utilized a cylindrical or coin cell design, in which the active component of the battery was supported inside a rigid metal casing. The flat-shaped batteries helped to reduce the total volume occupied by the battery and allowed for easy integration of the battery with the photographic films. The battery pack consisted of four cells connected in series to increase the voltage of the battery to 6.0 V. The fabrication process started by coating a vinyl film with a conductive material to serve as the current collector for the anode and cathode. Slurries of MnO_2 and Zn were deposited on the vinyl film with slot-die coating to serve as the cathode and anode of the battery, respectively. A cellophane film was placed between the anode and cathode to serve as the separator. The battery pack was encapsulated within aluminum casing to protect the battery. The theoretical capacity of the battery was around 270 mAh. Figure 6.5(b) shows the discharge characteristics of the battery at various discharge rates. While taking a photograph, the battery was discharged at a rate of 2 A for 50 ms. The battery pack was engineered to provide high pulses of power, which was necessary for powering the flash. The use of the flat battery packs was limited to the Polaroid cameras. In the late 2000s there was a reemergence of interest in flat printed primary batteries.

The availability of low-cost microelectronics, and advances in printed electronics led to the development of low-cost, disposable types of electronics for environmental sensing, activity tracking, and health monitoring purposes [28]. Primary batteries are more suitable than rechargeable batteries since these devices have a lifetime of a few days to a week. After the end of use, users can dispose of the sensor. Figure 6.6(a) shows a schematic of a printed battery laminated with a flexible electronic circuit. Figure 6.6(b) shows an example of a printed temperature sensor which can be placed on a human body to measure body temperature. In another example, a temperature sensor can be placed on temperature-sensitive produce or goods during shipping. The sensor can periodically transmit data to a mobile phone or control unit via a Bluetooth connection. In this type of sensor, a primary printed battery can be laminated with the sensor to power the electronics and communication. There is also considerable research effort into developing disposable tattoo-like sensors for measuring biological signals from human skin such as glucose and lactate levels during exercise (Figure 6.6(c)) [29]. The sensor works by measuring the concentration of glucose and lactate biomarkers in the sweat. If the level falls beyond a set limit, the user will be notified. Primary printed batteries are advantageous for these applications, since they are much safer in comparison to Li Ion batteries and they can be easily customized based on the requirements of the application.

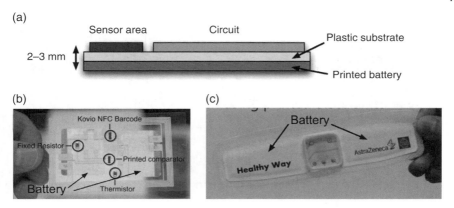

Figure 6.6 (a) Schematic of a printed battery laminated with a printed electronic device. (b) Demonstration of a printed temperature sensor with printed electronics powered by two-printed batteries connected in series. (c) Demonstration of printed batteries integrated with a hybrid sensor.

With the hope of taking advantage of this market, companies such as Enfucell and Blue Spark commercialized printed primary batteries in the late 2000s. The battery designs adopt conventional zinc-carbon chemistry. A conductive carbon paste is used to serve as the current collector for the anode and cathode. The impedance of these batteries is generally high (100 ohm) in comparison to conventional batteries (0.5 ohm) due to the low conductivity of the carbon electrode. These batteries are well suited for applications that required low amounts of power for extended periods of time. The typical voltage of a primary battery is around 1.5 to 1.65 V. The thickness of the battery ranges from 500 to 800 micron. Commercially available printed batteries are flexible to some extent due to their thin form factor but they are not designed for repeated flexing and twisting. Since the early 2010s there has been tremendous interest in academia in the design and fabrication of printed flexible primary batteries.

Pritesh *et al.* from the University of Cambridge in collaboration with Nokia Research Labs demonstrated one of the early designs of a printed flexible alkaline battery [30]. Figures 6.7(a) and 6.7(b) show the schematic and optical image of the battery, respectively. In this design, an electrospun carbon fiber electrode was used as the current collector for the MnO_2 electrode. During the printing process a part of the slurry was embedded inside the carbon fiber sheet, which improved the flexibility and adhesion of the electrode. A thin zinc sheet served as the anode in the battery. Due to the high conductivity of the zinc foil, no additional current collector was necessary. A dry TiO_2-doped PEO-based solid-state electrolyte was developed to serve as a separator between the electrodes. Figure 6.7(c) shows the discharge curves of the battery

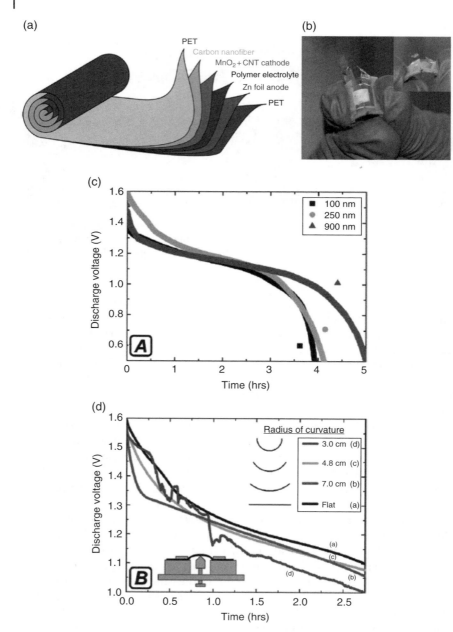

Figure 6.7 (a) Schematic of the Zn-MnO$_2$ flexible battery, (b) optical images of the battery, (c) discharge curves of the battery with varying diameter of carbon fiber and (d) discharge curves of the battery when flexed at various bending radii [30]. Copyright (2010), American Chemical Society.

with varying thickness of the carbon fiber. The ohmic resistance in the battery decreases with varying diameter of the carbon fiber. Figure 6.7(d) shows the performance of the battery when the battery was flexed to various bending radii. The battery was stable until a bending radius of 4 cm, beyond which cracking and delamination of the electrode occurred, which led to fluctuation of voltage during discharge. Delamination and cracking of the electrode are the most common modes of failure in printed batteries during flexing. Delamination increases ohmic losses within the battery, which decreases the energy density of the battery. Cracking increases particle-to-particle resistance and under certain conditions a part of the active layer can lose electrical contact with the rest of the layer, which reduces the capacity of the battery. The problem of delamination and cracking gets worse as the thickness of the electrode is increased or when the batteries are flexed to a lower bending radius.

Gaikwad *et al.* demonstrated a unique mesh-embedded electrode design, wherein the active layer of the battery was embedded inside a mesh [11]. Figure 6.8(a) shows a schematic of the process flow of creating the mesh-embedded electrodes. The mesh provides support and reinforcement for the active layers. With the mesh design the active layer can be completely folded without any delamination or cracking of the electrode. The fabrication process starts by dip coating the mesh with a slurry of the active material, followed by printing of a thin layer of silver ink over the electrodes to serve as the current collector of the electrodes. The printed current collector improves adhesion to the active layer and is more flexible in comparison to metal foils. Figures 6.8(b) and 6.8(c) show the discharge curves of the battery at various C-rates and the capacity retention of the battery when the battery is flexed at various bending radii. Figure 6.8(d) shows the battery with mesh-embedded electrodes powering a green-LED even after the battery is completely folded. A non-woven fabric can also be used in the place of a mesh. In both designs, the capacity of the battery depends on the thickness and the porosity of the support. The areal capacity and rate capabilities of the battery depend on the thickness of the support. Figures 6.8(e) and 6.8(f) show a flexible battery pack consisting of two cells stacked in series, which is integrated with an interactive strain sensor. When the sensor is flat the blue-LED display shows FLAT and when the sensor is flexed, the display shows BEND [31]. The flexible battery pack is used to power both the display and the microcontroller.

Electronics based on solution-processed organic transistors have been of interest since the early 1990s for application in flexible and large-area sensing [32–34]. Solution-based processing enables the fabrication of electronics from inks with the help of conventional high-speed printing technologies. Transistors are the basic building blocks of electronic circuits. Over the last 20 years, there has been tremendous progress in the design and synthesis of high mobility, oxygen-stable, n- and p-type organic semiconductors, which have reduced the operating voltages of transistors considerably [35]. However, in order to

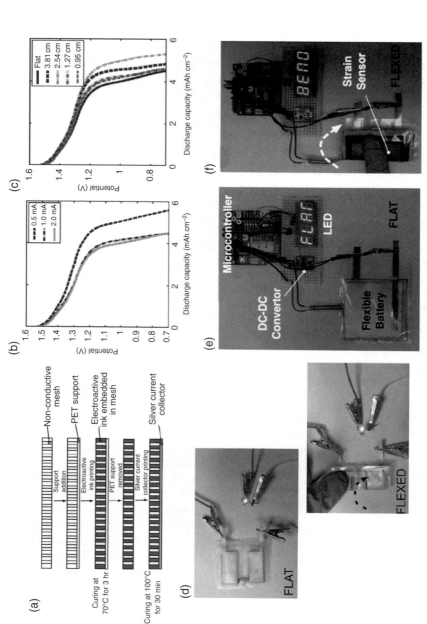

Figure 6.8 (a) Process flow of fabricating a mesh-embedded flexible alkaline battery. Discharge curves at the mesh-battery at various C-rates (b) and discharge curves of the battery when flexed to various bending radii (c). (d) Mesh-battery powering a green-LED under flat and bend conditions [11]. (e) A flexible alkaline battery with reinforced electrode structure powering an interactive display and microcontroller. The display shows FLAT when the battery is flat and the display shows BEND (f) when the battery is bent [31] Copyright (2012), John Wiley and Sons

improve the device yield and decrease device-to-device variability, the channel lengths and dielectric thicknesses in most printed circuits are still very large. The potential of a battery ranges from 1.5 to 4.2 V, which is lower than the voltage required for powering printed circuits. A DC-DC converter can increase the output voltage from the battery to the desired levels, but DC-DC converters are often inefficient and require the addition of more components into the system, and are therefore not recommended. It would be beneficial to fabricate a high voltage printed battery pack which can be laminated with the organic circuit (Figure 6.9(a)). Gaikwad *et al.* demonstrated a high voltage Zn-MnO$_2$ flexible battery where ten Zn-MnO$_2$ cells were printed on a porous substrate and connected in series to increase the voltage of the battery [18]. The schematic and photograph of the battery are shown in Figure 6.9(b). In this design, a cellulose-based alkaline battery separator was used as a support for the electrodes. The Zn and MnO$_2$ slurries were printed on the separator

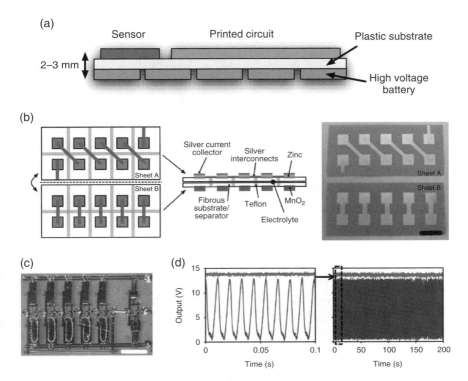

Figure 6.9 (a) Schematic showing the integration of a high voltage battery with a printed sensor. (b) Schematic of an all-printed high voltage Zn-MnO$_2$ battery. The battery consists of ten Zn-MnO$_2$ cells connected in series. (c) Optical image of a printed ring oscillator and (d) output from a five-stage ring oscillator powered with a high voltage printed battery [18]. Copyright (2013), AIP Publishing AIP.

with stencil printing. Silver ink was spray printed on the electrodes to form the current collectors. A hydrophobic amorphous Teflon dispersion was dispenser-printed between the electrodes to serve as a hydrophobic barrier for the aqueous electrolyte and prevent ionic contact between neighboring cells. The cells were connected in series by spray printing silver interconnects between the cells. After each printing step, the substrate was heated in an oven to dry the ink. A polymer gel electrolyte of KOH and ZnO was dispenser-printed on the electrodes and the substrate was folded along the dotted line to activate the battery. The design demonstrated by Gaikwad *et al.* can be easily adopted in a roll-to-roll fabrication process. Since all the cells of the battery are fabricated on a single substrate, the fabrication process was greatly simplified. The use of a printed Teflon barrier for the electrolyte reduced the dead space related to encapsulating each cell separately. The high voltage flexible battery was used to power an all-printed five-stage ring oscillator. The oscillator was fabricated with an inkjet printing process. Figure 6.9(c) shows the optical micrograph of the printed oscillator. The battery was used to power the oscillator for a period of 15 minutes (Figure 6.9(d)). With an input voltage of 14 V, an oscillatory output with a frequency of 100 Hz was observed. During the period of 15 minutes, the voltage of the battery remained the same since a very low current was required to drive the circuit.

At this point, primary printed battery technology is mature. In the last couple of years there have been numerous demonstrations by printed battery startup companies on the possible applications of disposable printed batteries. Products such as temperature sensors, smart tags, and golf swing trackers were demonstrated in collaboration with electronics companies. Currently there still is not a product available in the market that has achieved commercial success. The future of primary printed batteries will depend primarily on the development of new products and devices that leverage their unique characteristics.

6.3 Printed Rechargeable Batteries

The tremendous interest in wrist-worn health and fitness trackers and smart watches drew early efforts to develop printed rechargeable batteries as a potential power source to replace rigid Li Ion batteries [1, 28, 36]. Presently, the size of wearable devices is dictated by the size of the battery. It is not uncommon to see the battery taking up more than half the volume of the device. The availability of a high-performance, flexible, rechargeable printed battery opens up the opportunity to embed a battery within the strap of the watch. Such a design can help to reduce the overall thickness of the smart watch. Flexible batteries can also enable new product designs which were not possible previously. Other potential applications of rechargeable printed batteries include smart bankcards, smart labels, foldable devices, smart clothing, and tattoo-like sensors.

The first activity trackers were available in the market in the late 2000s. These devices were clip-on devices, which could be attached to clothing or embedded within shoes. A typical activity tracker contains a 3-axis accelerometer to track the movement of the user, gyroscope to measure orientation and rotation, microcontroller to process signals, rechargeable Li Ion battery to power the electronics, and Bluetooth to transmit the data. The raw data from the accelerometer and gyroscope are analyzed with an algorithm and converted to useful data such as the number of steps taken, total active time and calories burned. A simple activity tracker with a small LED display has a power consumption of around 5 to 15 mAh/day. Most activity trackers are powered with a small pouch-type Li Ion battery with a capacity in the range 30 to 70 mAh. On a single charge, the tracker can run for a period of 5 to 7 days. The clip-on activity trackers were soon replaced with wrist-worn activity trackers and smart watches. The barrier to adoption for the wrist-based devices was low since consumers were accustomed to wearing watches on their wrist. The first smart watches were introduced in the market in 2012. In addition to tracking activity levels and having Bluetooth connection, a smart watch contains a large color display, pulse sensor, GPS, and a media player. Smart watches are powered by Li Ion batteries with capacities in the range of 250 to 400 mAh. Smart watches typically last for a day on a single charge. The excessive use of the GPS function or the media player can drain the battery completely in a couple of hours. Nowadays, many smart watches have fast-charging capabilities, whereby the battery can be charged to 70–80% of its capacity within 30 minutes. Electronics companies such as Samsung, LG, and Philips have shown numerous demonstrations of futuristic, completely flexible devices such as bendable phones, displays, and tattoo-like health sensors. Excitement around these products galvanized research efforts into printed rechargeable batteries. Fully flexible devices are still in the early stages of development and are not expected to be available commercially for the next couple of years. In the meantime, wrist-worn devices are expected to be the primary consumers of printed rechargeable batteries. Currently, the electronics in smart watches are integrated on a rigid board with a footprint of a couple of cm^2 and a soft elastomeric strap is used to hold the sensor around the wrist. The straps have a total area in the range of 5–7 $inch^2$ and a thickness of 3–5 mm. Based on the available volume, a thin flexible printed battery with a capacity in the range around 50–100 mAh can be easily embedded inside the strap [37]. These batteries can serve as a primary power source for the simple activity trackers, or as a secondary back-up power source for smart watches that require higher battery capacity. Improving total battery capacity is particularly important for high-end smart watches since they last for less than a day on a single charge. In addition, it may be possible to design a system wherein the straps are attached to the sensor with a magnetic holder. If the battery in the device dies, the user can replace the dead batteries with charged ones and continue using the device.

A printed battery, when worn around the wrist, will experience compressive and tensile stresses which can lead to delamination of the active layer from the current collector film, swelling of the active layer, or loss of contact between the separator film and the electrode. Printed batteries for bendable applications should be designed to reduce mechanical stresses and prevent mechanical failures. Researchers have demonstrated numerous innovative designs to prevent cracking and delamination of the active layers. Some of the novel designs will be discussed in this section.

Liangbing Hu *et al.* reported one of the early designs of a printed rechargeable Li Ion battery [38]. In this design, the metal current collector foil was replaced with a carbon nanotube film to improve the flexibility of the electrode. Figures 6.10(a) and 6.10b show the schematic of the fabrication process. The fabrication process starts by blade coating carbon nanotube (CNT) solution on a stainless steel film, followed by blade coating slurries of the active layer (lithium cobalt oxide (LCO) or lithium titanate oxide (LTO)) on top of the CNT film. The CNT active layer film is removed from the stainless steel foil by a transfer process in water. The anode and cathode films are attached on either side of a Xerox paper with a PVDF binder. The Xerox paper serves as a separator between the anode and cathode, and a support for the electrodes. Figure 6.10(c) shows a photograph of the battery after the lamination process. Figure 6.10(d) shows a cross-section SEM micrograph of the electrode showing the interlayer between the current collector, active layer, and separator film. Figure 6.10(e) shows the charge–discharge characteristics of the battery. The battery retained its electrochemical performance even after repeated flexing cycles. Figure 6.10(f) shows a photograph of the battery powering a red-LED even under flexed state. This was one of the early demonstrations of using carbon nanotubes in a battery to improve its flexibility.

Carbon-based current collector electrodes have a lower conductivity in comparison to metal foils, which can lead to ohmic potential drops at high C-rates. To reduce ohmic potential drops, the conductivity of the flexible current collector films should be comparable to metal foils. Towards this goal, Lee *et al.* demonstrated a printed Li Ion battery where a nickel-coated textile film served as the current collector for the battery [39]. Figure 6.11(a) shows a schematic of the battery with nickel-coated textile as the current collector. Figure 6.11(c) shows the SEM micrograph of the nickel-coated textile fabric. The nickel is deposited on the textile fabric with an electroless deposition process. The design takes advantage of the fibrous structure of the fabric. A part of the active layers gets embedded within the fabric, which improves the adhesion between the current collector and active layer. The electrodes were highly flexible and no delamination or cracking occurred even after repeated folding and unfolding cycles. Figure 6.11(b) shows the textile after depositing of the metal and the slurry. Figures 6.11(d) and 6.11(e) show a comparison between a battery with the textile current collector and one with conventional metal foil, respectively.

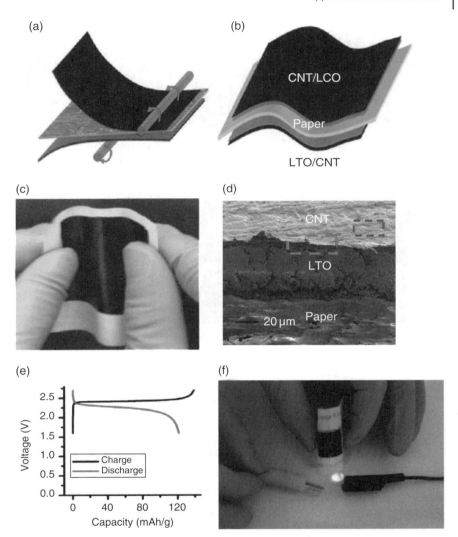

Figure 6.10 (a, b) Schematic of the flexible lithium ion battery with carbon nanotube-based current collector. The electrodes are attached to a Xerox paper with PVDF binder. (c) Photograph of the battery under flexed state, (d) cross-section SEM micrograph of the battery, (e) charge–discharge curve of the battery and (f) photograph of the battery powering a red-LED under flexed state [38]. Copyright (2010), American Chemical Society.

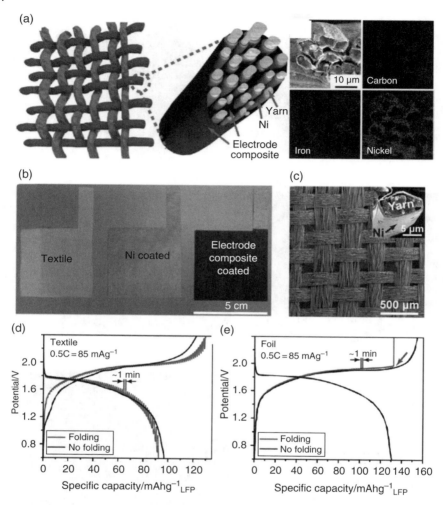

Figure 6.11 (a) Schematic of the nickel-coated textile fabric and EDS mapping showing the distribution of carbon, nickel, and iron. (b) Optical image showing the plain textile before and after electroless deposition of nickel and deposition of the slurry. (c) SEM micrograph showing the polyester textile with the nickel metal on the yarn. (d) and (e) Electrochemical characterization of the LFP- and LTO-based batteries when folding and unfolding the batteries with textile and metal foils as the current collector [39]. Copyright (2013), American Chemical Society.

(a)   (b)

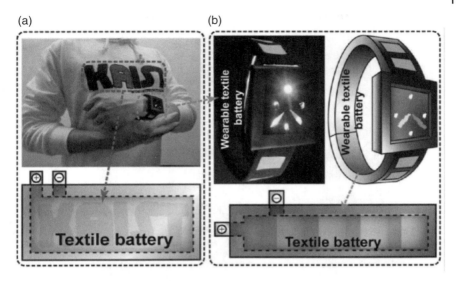

Figure 6.12 Demonstration of the textile battery attached to a shirt (a) and watch (b) [39]. Copyright (2013), American Chemical Society.

The battery with standard current collector foil failed after repeated flexing, while the textile battery retained its capacity. The textile batteries are suitable for powering watches and smart clothing. Figures 6.12(a) and 6.12(b) show the integration of the textile battery with clothing and smart watches.

The authors used this technology to develop large-area flexible battery modules with output voltages of 7.5 and 30 V by integrating multiple batteries in series [40]. Figure 6.13(a) shows an illustration of a high voltage battery module, which can be rolled up when not in use. Figure 6.13(b) shows details of the battery module. The performance of the battery module was comparable to the performance of individual cells. High voltage flexible battery modules are useful for powering applications that require high driving voltages. Figure 6.13(c) shows the integration of the high voltage textile battery module with a tent. Figure 6.13(d) shows the integration of the battery module with window binder. The battery module can be used to store energy generated by flexible solar cells.

The use of thin metal foils in printed flexible batteries has been limited due to the sharp edges of the foil, which are likely to penetrate through the separator and create a short. Choi *et al.* solved this problem by developing a coplanar battery architecture [41]. Figures 6.14(a) and 14(b) show an optical image and schematic of the interdigitated battery. In this design, the slurries for the anode and cathode are deposited on conventional metal foils. The electrodes are then cut into fork-shaped patterns with a custom-designed knife. The anode and

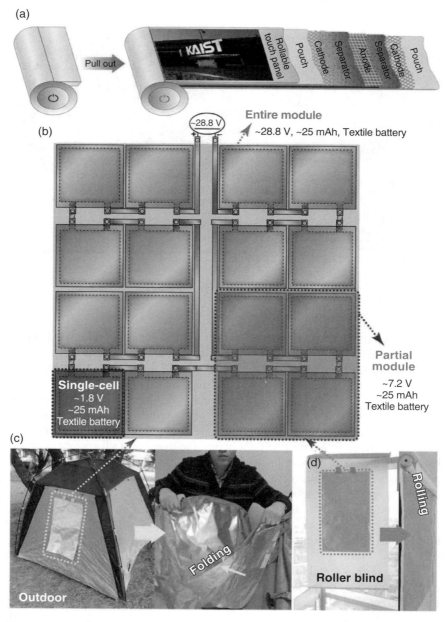

Figure 6.13 (a) Illustration of the unrolling the textile-based flexible lithium ion battery. (b) Schematic of the battery module containing 16 batteries in series to increase the potential to 29 V with a capacity of 25 mAh. Photographs of the textile battery module integrated in a tent (c) and roller blind (d) [40]. Copyright (2014), The Royal Society of Chemistry.

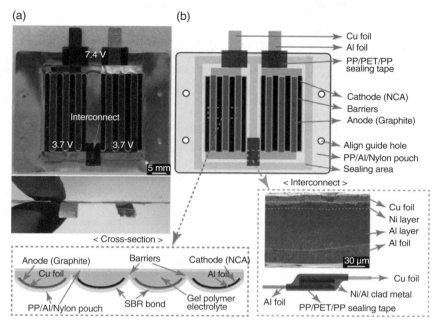

Figure 6.14 Optical image (a) and schematic illustration (b) of the coplanar flexible battery. The battery is composed of two cells connected in series. The electrodes have an interdigitated design. The cross-section at the bottom shows the curved shape of the current collector foil. The curved design helps to prevent shorting of the electrode during flexing [41]. Copyright (2015), American Chemical Society.

cathode electrodes are then placed in an interdigitated design and encapsulated within metal-laminated pouches. Two batteries are connected in series to form a flexible battery module with an output voltage of 7.4 V. This design helps to prevent shorting of the battery during flexing. The authors demonstrated numerous applications of the coplanar batteries. Due to their thin form factor, these batteries are suitable for smart card applications. Figures 6.15(a), 6.15(b) and 6.15(c) show the integration of the battery with a wrinkle-smoothing cosmetic patch, smart bank cards and smart watches, respectively.

For many wearable applications such as smart clothing, tattoo-like sensors and health trackers, in addition to flexibility, stretchability is necessary to ensure conformal contact between the device and body [8, 42, 43]. Stretchable batteries are difficult to design and in the past couple of years there have only been a handful of stretchable batteries reported in the literature [12, 44–46]. Kettlgruber *et al.* demonstrated a rechargeable stretchable Zn-MnO$_2$ battery where the active layers were embedded inside a nickel mesh [47]. Figure 6.16(a) shows a schematic of the stretchable battery. An acrylic elastomer served as a support

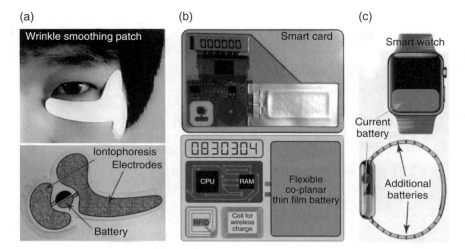

Figure 6.15 Examples of devices powered with a flexible coplanar battery. (a) A cosmetic patch with iontophoresis function powered with a flexible battery, (b) smart bank card to provide additional security for transactions powered with a battery, and (c) smart watch with flexible batteries embedded inside the band of the watch to provide additional capacity [41]. Copyright (2015), American Chemical Society. (*See insert for color representation of the figure.*)

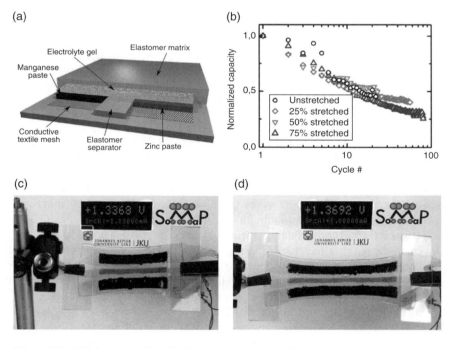

Figure 6.16 (a) Schematic of Zn-MnO$_2$ stretchable battery, (b) discharge capacities of the stretchable battery with cycle number with the battery held at various state of stretching, and (c, d) demonstration of the battery retaining its voltage even after stretching of the battery by 50% [47]. Copyright (2013), The Royal Society of Chemistry.

and encapsulation for the battery. The battery retained its capacity even after stretching the battery to a strain of 75% (Figure 6.16(b)). The accessible capacity of the battery dropped by 50% after 100 electrochemical cycles due to irreversible reaction on the cathode side. Figure 6.16(c) and 6.16(d) show optical images of the battery before and after stretching, respectively. The voltage of the battery remained constant even after stretching. The group of Prof. Hanqing Jiang from Arizona State University demonstrated a novel battery design based on the ancient Chinese art of kirigami [48, 49]. In this design, thin metal foils are cut in predetermined patterns which allow them to stretch and twist, in-plane or out-of-plane. Once the current collector is cut, the slurry of cathode and anode are printed over the current collector foils and dried. The electrodes are stacked together with a thin porous separator between the anode and cathode, and encapsulated within metal-laminated pouches. Figures 6.17(a) and 6.17(b) show optical images of the kirigami battery under the compressed and most stretched state. The kirigami battery was able to continuously power a Samsung Galaxy Gear smart watch even after stretching of the battery (Figures 6.17(c) and 6.17(d)). Figures 6.17(e) and 6.17(f) show discharge curves of the kirigami battery discharged at 2.9 and 48 mA, respectively, representing the standby mode and watching a video on the watch.

Printed rechargeable Li Ion batteries are going to play an important role in the success of flexible electronics. Printed Li Ion batteries benefit directly from the developments in rigid Li Ion batteries because new materials, electrode designs, and additives can be easily applied to the printed Li Ion batteries. Currently the areal capacities of printed Li Ion batteries are much lower than those of conventional batteries. The development of new battery designs described in the following section can increase the areal capacities of the batteries and make them more attractive for commercial products.

6.4   High-Performance Printed Structured Batteries

The high energy density and long cycle life of Li Ion batteries have made them the primary choice of battery chemistry for powering consumer electronics and electric vehicles (EV). The conventional Li Ion battery adopts a 'jellyroll' design, in which thin foils of copper and aluminum serve as the current collector for the anode and cathode, respectively. Slurries of the anode and cathode material are deposited on the current collector foils with the slot-die printing method. The anode, separator, and cathode layers are stacked together and winded to form a 'jellyroll' [50]. The jellyroll is soaked with an electrolyte solution containing lithium salt and encapsulated under compression within metal-laminated pouches [13]. Even though the basic design of an Li Ion battery has remained the same over the last 25 years, a year-by-year increase in energy density of around 5 to 10% has been consistently achieved by the

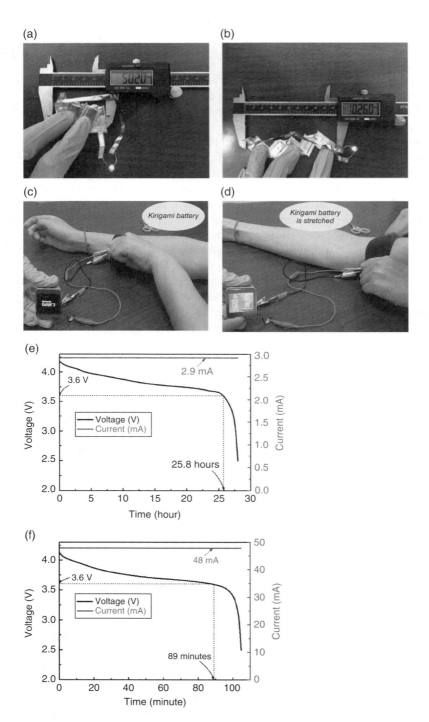

Figure 6.17 Photograph of the Kirigami battery in compressed (a) and stretched (b) state. (c, d) Photograph of the Kirigami battery continuously powering a Samsung Galaxy Gear watch even under stretched state. Discharge curves at a slow rate (e) representing standby mode and at a high rate (f) representing watching a video on the watch [48].

shrinking of the thickness of the inactive components, and incremental improvements in the material design. The thickness of the current collector and separator films in the early generations of Li Ion batteries were around 30 to 50 micron. In the current state-of-the-art Li Ion batteries, the thickness of the current collector and polymer separator films is as low as 5 micron. The transition from metal casings to metal-laminated pouches in the early 2010s also played an important role in improving the energy density of Li Ion batteries. The pouch architecture allows for custom designs based on the power requirements of the device. Many mobile products such as cell phones, tablets, and laptops have shifted to non-replaceable pouch batteries in recent years, in which the battery is fixed to the circuit board. This design helps to reduce the overall volume of the product. Battery packs are quickly reaching the limits of practical energy density possible by the engineering route. A leap from the current technology will require new battery designs and advances in novel chemistries such as magnesium-ion, lithium-air and lithium-sulfur batteries. New battery technologies generally take a very long time to develop and commercialize. In the meantime, battery engineers are investigating novel printing technologies as a tool to fabricate structured battery electrodes to improve the rate performance and energy density of Li Ion batteries.

A battery electrode is composed of electrochemically active materials mixed with carbon-based conductive additive, and a polymeric binder to hold the electrode together. After the slurry is deposited on the current collector foil and dried, a porous structure is formed. The electrodes are calendared (compressed) to a porosity of 20–30% to increase their density and particle-to-particle contact [51–54]. The percolative network of pores between the solid particles is filled with an electrolyte solution. The energy density of the batteries can be improved by further increasing the density of the active layers or the thickness of the electrodes. As the porosity of the electrode is reduced, the tortuosity increases, which increases the distance the lithium ions have to travel as they move between the anode and cathode. For battery electrodes with fast reaction kinetics and high electrical conductivity, the rate performance is limited by the diffusion of lithium ions through the electrolyte [55]. The increase in tortuosity of the electrode will increase polarization losses at high rates. On the other hand, increasing the thickness of the electrodes can increase the fraction of the volume occupied by the active materials in the battery. But electrodes beyond 100 micron are mechanically weak, which can reduce the calendaring life of the battery. The effective tortuosity of the battery also increases with the thickness of the electrode, which degrades the rate performance of the battery. Hence increasing the density or thickness of the electrode may not be a viable option for increasing the energy density of the battery, in particularly for applications where high discharge rates are required.

Researchers are actively looking into printing technology as a tool to fabricate battery electrodes with structured designs [2, 8, 16, 19]. In a conventional

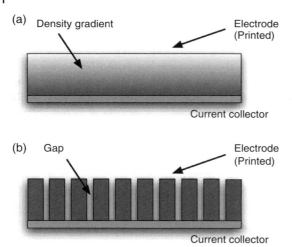

Figure 6.18 (a) Schematic of structured electrode with density gradient along its thickness. The electrode near the current collector is denser as compared to the electrode away from the current collector. (b) Schematic of structured electrode with gaps between islands of active material.

battery electrode, the density and porosity of the active layer is homogeneous throughout the battery. Additive printing technology opens up the possibility of fabricating electrodes with a gradient of porosity and conductivity across its thickness. Such a design can help to improve the energy density and rate capabilities of the batteries. In an Li Ion battery, the reaction boundary during charge and discharge processes starts at the interface near the separator and then moves toward the current collector. At high C-rates, a part of the electrode near the current collector does not take part in the reactions due to slow diffusion of Li Ion through the electrolyte phase. By using printing technologies, a structured electrode can be fabricated to overcome the diffusion limitations and increase utilization of the electrode. For example, Figure 6.18(a) shows design of an electrode in which the density of electrode changes along its thickness such that the electrode near the separator is more porous in comparison to the material near the current collector. Such a design can help with the transport of Li Ion between the electrodes and increase utilization at high rates. A similar concept can also be used to vary the electrical conductivity across the electrode such that the interface near the separator is more conductive than the region close to the current collector, to ensure that current can be easily passed and removed from regions away from the current collector. The porosity or conductivity variation in the electrode can be achieved by printing multiple layers of slurries with different formulations on top of each other. The conductivity and porosity of the electrode can be changed by varying the weight of carbon additive and solvent content. Printed structured electrodes can also be used to improve the mechanical flexibility of batteries. Repeated flexing leads to swelling of the electrodes, which can lead to a loss in particle-to-particle contact and delamination of the electrodes. The mechanical degradation due to flexing can be prevented by printing islands of the active material on the foils rather than printing a homogenous layer (Figure 6.18(b)). The gap between the islands can serve as a buffer to absorb the stresses during flexing. The gap can also help to increase the rate capability of the battery by providing pathways for rapid diffusion of Li Ion through the electrodes.

Researchers at Palo Alto Research Center (PARC) developed a co-extrusion method to deposit high aspect ratio electrodes at print speeds of 20 cm/s [19, 56]. This technology was initially developed to deposit high aspect ratio silver electrodes to serve as interconnects in silicon solar cells. The printer head was later modified to print battery slurries. Figure 6.19(a) shows a schematic of the co-extrusion print head. The printer was designed to print thick electrodes with

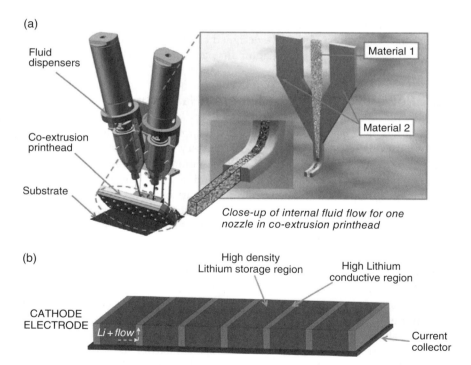

Figure 6.19 (a) Schematic of the co-extrusion printing process developed by researchers at Palo Alto Research Center. (b) Schematic of structured electrode with alternating patterns of high-density and low-density regions to improve Li Ion diffusion through the electrode [56]. Copyright (2014), Elsevier.

alternating strips of high- and low-density regions (Figure 6.19(b)). The high-density regions are energy dense and store the bulk of the energy of the battery. The narrow low-density regions have low tortuosity, and they provide pathways for fast Li Ion diffusion through the electrode. With the multi-porosity design, the thickness and the energy density of the electrodes can be increased without affecting the rate performance of the batteries. Using this concept, the volumetric energy density of the battery can be increased by 10% by increasing the thickness of the electrodes without affecting rate performance.

There are a number of challenges battery engineers must solve before structured electrodes can be used in commercial batteries. The primary challenge is the cost of the battery. Presently the money and time cost of printed structured electrodes is high in comparison to that of slot-die printed electrodes. Any increase in production time will add to the cost of the battery. Batteries with structured electrodes should present at least a 10–15% improvement in energy density before they will be economically viable for adoption by battery manufacturers. The calendaring life of the battery is closely related to the mechanical strength of the electrodes. The strength of the electrode decreases exponentially as the thickness of the electrode is increased beyond a limit of 60 to 80 microns. The battery should be designed carefully based on the application to prevent any mechanical degradation during use.

6.5 Power Electronics and Energy Harvesting

When designing a battery system for a given application, one essential choice is whether to use a primary or secondary battery. Primary batteries based on stable and low-cost alkaline chemistries are of great interest for powering disposable electronics, such as smart labels and some medical devices. Many reusable consumer electronics also rely on primary batteries with standardized dimensions, such as coin cells and AA cells, which must be replaced periodically by the user. Replacing a printed and flexible battery is less straightforward since standardized flexible battery shapes do not yet exist. Therefore, if the desired product life is longer than the battery life, it is preferable to use a secondary (rechargeable) battery instead.

To charge secondary batteries, many options are available. Wired charging, for example, using the universal serial bus (USB) standard, is commonly used for batteries in consumer electronics. However, some systems, such as flexible or stretchable wearable electronics and electronics which may be exposed to water, cannot tolerate bulky USB connectors or exposed metal. The batteries in these systems can still be charged on demand using wireless methods such as inductive charging or radio frequency (RF) power transmission. Batteries can also be charged with energy-harvesting devices that collect energy from ambient sources such as light, heat, and motion. Energy harvesting is especially

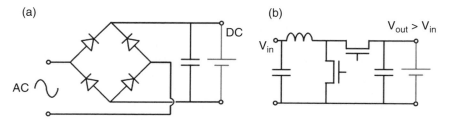

Figure 6.20 Power electronics commonly used in battery charging. (a) The bridge rectifier circuit converts from AC to DC power. (b) The boost converter converts from a lower DC voltage to a higher DC voltage. The two transistors in the boost converter are switched on and off at opposite times; the conversion ratio V_{in}/V_{out} is controlled by the duty cycle of the switching.

important for applications such as sensor networks with large numbers of nodes or devices in hard-to-access locations which must operate for long periods of time without any user involvement.

Since battery charging requires direct current (DC) and a particular voltage range (dependent on the battery chemistry), power electronics are needed when these requirements differ from the output characteristics of the energy source. If the energy source produces alternating current (AC), for example, a rectifier is necessary. Figure 6.20(a) shows a common rectifier circuit, which uses a diode bridge to rectify the AC signal and a capacitor to smooth the DC output. If the source produces DC power at a different voltage from the battery, a DC-DC converter such as the boost converter shown in Figure 6.20(b) is needed. This circuit uses a combination of passive components and two transistors switched alternately to convert from a lower voltage to a higher voltage. A similar circuit with the input and output reversed can be used to convert from a higher to a lower voltage. For sources such as solar cells, which produce maximum power at a particular voltage, DC-DC converters known as maximum power point trackers (MPPT) are often used to ensure the device is operating at that voltage regardless of the voltage of the battery or load. DC-DC converters can also be connected between the battery and the load, in systems where the load requires a well-regulated voltage or a different voltage than that which the battery produces.

Battery management systems (BMS) are also very important to maximize the performance and safety of batteries. One critical function of a BMS is to terminate charging when the battery reaches its maximum voltage, preventing over-voltage from occurring. This can be accomplished using a transistor in series between the energy source and the battery or, for current-limited sources, a transistor or diode in parallel with the battery to shunt away excess current when the battery is charged. Li Ion battery management systems in particular often enforce a constant-current constant-voltage charging profile, charging the battery with a constant current until it reaches its maximum voltage, then maintaining that voltage and allowing the current to decay to zero. The constant-current portion charges the battery most of the way at a high rate, and the constant-voltage portion ensures the battery receives a complete charge without exceeding its voltage limit. Another important function of a BMS is to disconnect the load from the battery when the battery reaches its minimum voltage, corresponding to the discharged state. Some BMSs also monitor the battery temperature and adjust the charging and discharging conditions appropriately.

Typically, these circuits are produced by soldering surface-mount components to a circuit board, but there is growing interest in developing printed and flexible components for power electronics. Numerous printed and flexible diodes have been reported for rectifier applications. Printed inductors and capacitors have been optimized for use in DC-DC converters and have shown performance competitive with their surface-mount counterparts. Flexible thin film transistors (TFTs) have also been employed in a number of power electronic circuits although further research is required for the TFT performance to be competitive with silicon transistors for these applications.

Among energy-harvesting sources, photovoltaics (PV), which generate electricity from light, are particularly appealing because the available power density from sunlight and even indoor light is relatively large. Each PV cell produces a voltage of the order of several tenths of a volt, which varies roughly logarithmically with illumination. The PV current varies roughly linearly with illumination, ranging from tens of $\mu A/cm^2$ under indoor lighting up to tens of mA/cm^2 in full sun. To charge a battery, multiple PV cells are usually connected in series to reach the battery voltage. Flexible amorphous silicon PV modules are commercially available and have therefore been utilized alongside batteries in a number of energy systems. Figures 6.21(a) and (b) show a flexible energy system combining such a module with a printed Li Ion battery in a layered thin film structure [37]. The particular PV module was selected because it produces its maximum power output when operated over the battery voltage range, under both indoor and outdoor illumination conditions, as shown in Figure 6.21(c)–(d), allowing the system to operate efficiently without MPPT. There is also a great effort toward using solution-processable materials such as organics and perovskites to fabricate low-cost printed PV modules. For example, we have developed the printed organic PV modules shown in Figure 6.22(a) and integrated them with a BMS and a printed Li Ion battery as shown in Figure 6.22(b). Under sunlight, the solar module charges the battery as shown in Figure 6.22(c), with a current of approximately 2 mA.

Since many applications of printed batteries involve moving objects, harvesting energy from motion is also of great interest. For example, energy can be

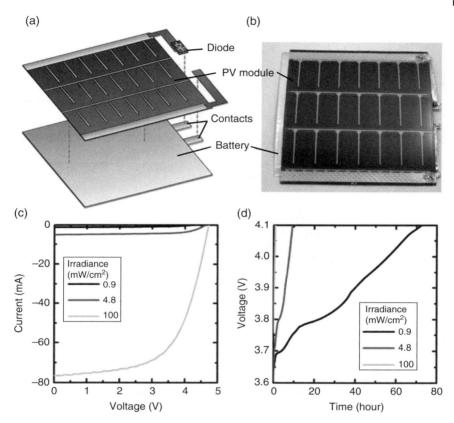

Figure 6.21 Flexible thin film system using PV module and Li Ion battery. (a) Illustration and (b) photograph of the system. (c) Current-voltage characteristics of the PV module under three illumination conditions, corresponding to sunlight (green), and indoor light with high brightness (red) and moderate brightness (blue). (d) Battery voltage curves when charged by the PV module under the same illumination conditions [37]. Copyright (2016), Nature Publishing Group.

harvested from body motions such as footsteps or movement of joints and used to power wearable health sensors. Similarly, energy harvested from vibrations of industrial and manufacturing equipment can be used to power condition-monitoring sensors. Two leading methods of harvesting energy from movement are piezoelectrics and triboelectrics. Piezoelectric materials change their polarization when subjected to an external force, thus creating an electric potential in one direction under tension and in the other direction under

(a)   (b)

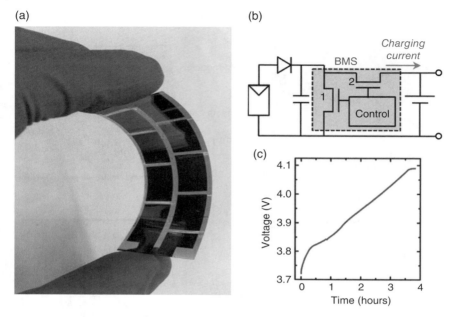

Figure 6.22 Battery charging using printed organic photovoltaic module. (a) Photograph of the photovoltaic module demonstrating its flexibility. (b) Circuit schematic of battery charging. The components in the blue shaded box are contained in a BMS integrated circuit. (c) Voltage profile of a printed Li Ion battery during solar charging. The BMS prevents the battery voltage from increasing beyond 4.1 V, by turning on the transistor labeled "1" to shunt away excess current from the PV module, around the 3.6-hour mark.

compression. A triboelectric generator, on the other hand, employs two materials, which experience electron transfer when brought into contact. An example of a triboelectric generator is shown in Figure 6.23(a), in which nickel-coated fibers and parylene-coated fibers are woven into a textile. Contact between the fibers induces electron transfer from the nickel (blue) to the parylene (yellow); when the materials are separated again the electrons flow in the opposite direction if there is a pathway through an external circuit. Using this triboelectric generator, energy from the motion of the arm past the side of the body was collected to charge a flexible Li Ion battery integrated into a belt, and to power a wearable heartbeat monitor as shown in Figure 6.23(b)–(e). Since most motions produce an AC signal, a rectifier such as that shown in Figures 6.20(a) and 6.23(d) is needed for battery-charging applications. These generators tend to produce high voltages, often many volts at open circuit, but relatively low currents. Therefore, multiple generators are often connected in parallel to increase the charging rate.

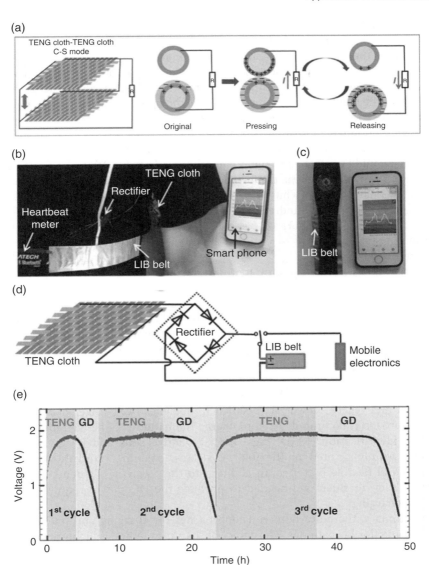

Figure 6.23 Wearable system with triboelectric generator and Li Ion battery. (a) Illustration of the working mechanism of a textile triboelectric generator using fibers coated with nickel (blue) and parylene (yellow) in contact-separation (C-S) mode. (b–c) Photographs of wearable system including textile triboelectric nanogenerator (TENG) and belt containing flexible Li Ion batteries (LIB), used to power a heartbeat meter and transmit heart rate data to a smartphone. (d) Circuit schematic of the system. (e) Voltage profiles of the Li Ion battery during three cycles of charging by the triboelectric nanogenerator and galvanostatic discharging (GD) [23]. Copyright (2016), John Wiley and Sons.

Human bodies and operating machines also produce heat, which can be harvested using thermoelectric generators. Both the voltage and the current produced by a thermoelectric generator are proportional to the temperature difference the device experiences. To harvest a large amount of power, therefore, the heat source must have a high temperature and be in good contact with one side of the thermoelectric generator, while the opposite side of the generator must be cooled sufficiently. Thermoelectric generators have been developed with a number of form factors, including wearable textiles [57–60] and patches for harvesting body heat, as well as a variety of rigid and flexible structures that could be attached to pipes and other equipment. While the flexible generators provide excellent conformability to surfaces, they often produce only a few millivolts each. Thus, reaching a voltage adequate for battery charging requires connecting hundreds of generators in series or employing a step-up DC-DC converter. Figure 6.24 shows an example thermoelectric battery charging system, consisting of a printed thermoelectric generator, printed battery, and DC-DC converter.

Controllable wireless charging methods such as RF and inductive power transfer are advantageous when ambient energy sources are unreliable or insufficient to meet a system's demands. In inductive charging, a receiving coil in the same package as the battery is brought close (typically within a few millimeters) to a transmitting coil. The magnetic field caused by an alternating current in the transmitting coil induces a current in the receiving coil, transferring energy. Inductive charging of printed and flexible batteries has been demonstrated using the widely accepted Qi standard as well as customized configurations. In RF power transfer, resonant circuits are used to transmit and receive power at specified frequencies over longer distances. This approach is commonly used to charge the batteries in active radio frequency identification (RFID) tags. Both inductive coils and RF antennas can be readily printed on flexible substrates, enabling integration with printed and flexible batteries. In addition to the receiving elements themselves, diode rectifiers are needed to produce the DC power required for battery charging.

In summary, there are numerous approaches for charging and managing printed and flexible batteries. The choice of charging method is highly dependent upon the details of the application, such as the load power consumption, desired charging frequency, required form factor, and power density available from the various ambient sources. Power electronics, which must be chosen appropriately for the energy source, are critical to ensure efficient, safe, and reliable battery operation. Finally, novel power electronic circuit topologies and physical integration schemes have been developed in order to charge batteries using multiple energy sources, enabling reliable and multifunctional power systems.

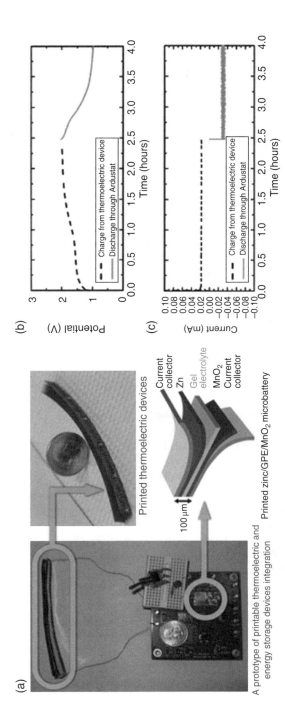

Figure 6.24 Printed thermoelectric device charging a printed battery. (a) Photograph of the system, consisting of thermoelectric generator, battery, and DC-DC converter. (b) Voltage and (c) current as the battery is charged by the thermoelectric device and discharged through an Ardustat [21]. Copyright (2012), IOP Publishing.

References

1 Gaikwad, A.M., Arias, A.C., Steingart, D.A. (2015) *Energy Technol.* **3**, 305.
2 Steingart, D., Ho, C.C., Salminen, J., Evans, J.W., Wright, P.K. (2007) *6th Int. IEEE Conf. Polym. Adhes. Microelectron. Photonics, Polytronic 2007, Proc.*, 261.
3 Ho, C.C., Evans, J.W., Wright, P.K. (2010) *J. Micromech. Microeng.* **104009.**
4 Braam, K.T., Volkman, S.K., Subramanian, V. (2012) *J. Power Sources* **199**, 367.
5 Braam, K., Subramanian, V. (2015) *Adv. Mater.* **27**, 689.
6 Gaikwad, A.M., Gallaway, J.W., Desai, D., Steingart, D.A. (2011) *J. Electrochem. Soc.* **158**, A154.
7 Long, J.W., Dunn, B., Rolison, D.R., White, H.S. (2004) *Chem. Rev.* **104**, 4463.
8 Ahn, B.Y., Duoss, E.B., Motala, M.J., Guo, X., Park, S.-I., Xiong, Y. *et al.* (2009) *Science* **323**, 1590.
9 Russo, A., Ahn, B.Y., Adams, J.J., Duoss, E.B., Bernhard, J.T., Lewis, J.A. (2011) *Adv. Mater.* **23**, 3426.
10 Aleeva, Y., Pignataro, B. (2012) *J. Mater. Chem. C* **2**, 6436.
11 Gaikwad, A.M., Whiting, G.L., Steingart, D.A., Arias, A.C. (2011) *Adv. Mater.* **23**, 3251.
12 Xu, S., Zhang, Y., Cho, J., Lee, J., Huang, X., Jia, L. *et al.* (2013) *Nat. Commun.* **4**, 1543.
13 Arthur, T.S., Bates, D.J., Cirigliano, N., Johnson, D.C., Malati, P., Mosby, J.M. *et al.* (2011) *MRS Bull.* **36**, 523.
14 Braam, K.T., Volkman, S.K., Subramanian, V. (2012) *J. Power Sources* **199**, 367.
15 Ho, C.C., Murata, K., Steingart, D.A., Evans, J.W., Wright, P.K. (2009) *J. Micromechanics Microengineering* **19**, 94013.
16 Sun, K., Wei, T.S., Ahn, B.Y., Seo, J.Y., Dillon, S.J., Lewis, J.A. (2013) *Adv. Mater.* **25**, 4539.
17 Bates, J.B., Dudney, N.J., Neudecker, B., Ueda, A., Evans, C.D. (2000) *Solid State Ionics* **135**, 33.
18 Gaikwad, A.M., Steingart, D.A., Nga Ng, T., Schwartz, D.E., Whiting, G.L. (2013) *Appl. Phys. Lett.* **102**, DOI 10.1063/1.4810974.
19 Cobb, C.L., Ho, C.C. (2016) *Electrochem. Soc. Interface* **25**, 75.
20 Madan, D., Wang, Z., Chen, A., Winslow, R., Wright, P.K., Evans, J.W. (2014) *Appl. Phys. Lett.* **104**, 2012.
21 Wang, Z., Chen, A., Winslow, R., Madan, D., Juang, R.C., Nill, M. *et al.* (2012) *J. Micromechanics Microengineering* **22**, 94001.
22 Fu, K., Yao, Y., Dai, J., Hu, L. (2016) *Adv. Mater.* **1603486.**
23 Fu, K., Wang, Y., Yan, C., Yao, Y., Chen, Y., Dai, J. *et al.* (2016) *Adv. Mater.* **28**, 2587.
24 Hu, J., Jiang, Y., Cui, S., Duan, Y., Liu, T., Guo, H. *et al.* (2016) *Adv. Energy Mater.* **6**, 1.

25 Golodnitsky, D., Yufit, V., Nathan, M., Shechtman, I., Ripenbein, T., Strauss, E. et al. (2006) *J. Power Sources* **153**, 281.
26 Chamran, F., Yeh, Y., Min, H.S., Dunn, B., Kim, C.J. (2007) *J. Microelectromechanical Syst.* **16**, 844.
27 Reddy, T. (2010) *Linden's Handbook of Batteries*, McGraw-Hill Education.
28 Kim, D.-H., Lu, N., Ma, R., Kim, Y.-S., Kim, R.-H., Wang, S. *et al.* (2011) *Science* **333**, 838.
29 Ko, H., Kapadia, R., Takei, K., Takahashi, T., Zhang, X., Javey, A. (2012) *Nanotechnology* **23**, 344001.
30 Hiralal, P., Imaizumi, S., Unalan, H.E., Matsumoto, H., Minagawa, M., Rouvala, M. *et al.* (2010) *ACS Nano* **4**, 2730.
31 Gaikwad, A.M., Chu, H.N., Qeraj, R., Zamarayeva, A.M., Steingart, D.A. (2013) *Energy Technol.* **1**, 177.
32 Arias, A.C., MacKenzie, J.D., McCulloch, I., Rivnay, J., Salleo, A. (2010) *Chem. Rev.* **110**, 3.
33 Arias, A.C., Ready, S.E., Lujan, R., Wong, W.S., Paul, K.E., Salleo, A., Chabinyc, M.L. (2004) *Appl. Physics Lett.* **85**, 3304.
34 Street, R.A., Wong, W.S., Ready, S.E., Chabinyc, M.L., Arias, A.C., Limb, S. et al. (2006) *Mater. Today* **9**, 32.
35 Dimitrakopoulos, C.D., Malenfant, P.R.L. (2002) *Adv. Mater. (Weinheim, Ger.)* **14**, 99.
36 Jung, S., Hong, S., Kim, J., Lee, S., Hyeon, T., Lee, M., Kim, D.-H. (2015) *Sci. Rep.* **5**, 17081.
37 Ostfeld, A.E., Gaikwad, A.M., Khan, Y., Arias, A.C. (2016) *Sci. Rep.* **6**, 26122.
38 Hu, L.B., Wu, H., La Mantia, F., Yang, Y.A., Cui, Y. (2010) *ACS Nano* **4**, 5843.
39 Lee, Y.H., Kim, J.S., Noh, J., Lee, I., Kim, H.J., Choi, S. *et al.* (2013) *Nano Lett.* **13**, 5753.
40 Kim, J.-S., Lee, Y.-H., Lee, I., Kim, T.-S., Ryou, M.-H., Choi, J.W. (2014) *J. Mater. Chem. A* **2**, 10862.
41 Kim, J.S., Ko, D., Yoo, D.J., Jung, D.S., Yavuz, C.T., Kim, N.I., Choi, I.S., Song, J.Y., Choi, J.W. (2015) *Nano Lett.* **15**, 2350.
42 Rogers, J.A., Someya, T., Huang, Y. (2010) *Science* **327**, 1603.
43 Lee, S., Kim, J., Jang, H., Yoon, S.C., Lee, C., Hong, B.H. *et al.* (2011) *Nano Lett.* **11**, 4642.
44 Kaltenbrunner, B.M., Kettlgruber, G., Siket, C., Bauer, S., Schwo, R. (2010) *Adv. Mater.* **22**, 2065.
45 Gaikwad, A.M., Zamarayeva, A.M., Rousseau, J., Chu, H., Derin, I., Steingart, D.A. (2012) *Adv. Mater.* **24**, 5071.
46 Wang, C., Zheng, W., Yue, Z., Too, C.O., Wallace, G.G. (2011) *Adv. Mater.* **23**, 3580.
47 Kettlgruber, G., Kaltenbrunner, M., Siket, C.M., Moser, R., Graz, I.M., Schwödiauer, R., *et al.* (2013) *J. Mater. Chem. A* **1**, 5505.

48 Song, Z., Wang, X., Lv, C., An, Y., Liang, M., Ma, T. et al. (2015) *Sci. Rep.* **5**, 10988.

49 Song, Z., Ma, T., Tang, R., Cheng, Q., Wang, X., Krishnaraju, D. *et al.* (2014) *Nat. Commun.* **5**, 1.

50 Marks, T., Trussler, S., Smith, A.J., Xiong, D., Dahn, J.R. (2011) *J. Electrochem. Soc.* **158**, A51.

51 Liu, G., Zheng, H., Simens, A.S., Minor, A.M., Song, X., Battaglia, V.S. (2007) *J. Electrochem. Soc.* **154**, A1129.

52 Zheng, H., Tan, L., Liu, G., Song, X., Battaglia, V.S. (2012) *J. Power Sources* **208**, 52.

53 Zheng, H., Liu, G., Song, X., Ridgway, P., Xun, S., Battaglia, V.S. (2010) *J. Electrochem. Soc.* **157**, A1060.

54 Liu, G., Zheng, H., Song, X., Battaglia, V.S. (2012) *J. Electrochem. Soc.* **159**, A214.

55 Nemani, V.P., Harris, S.J., Smith, K.C. (2015) *J. Electrochem. Soc.* **162**, A1415.

56 Cobb, C.L., Blanco, M., (2014) *J. Power Sources* **249**, 357.

57 Du, Y., Cai, K., Chen, S., Wang, H., Shen, S.Z., Donelson, R., Lin, T. (2015) *Sci. Rep.* **5**, 6411.

58 Kim, M.-K., Kim, M.-S., Lee, S., Kim, C., Kim, Y.-J. (2014) *Smart Mater. Struct.* **23**, 105002.

59 Dun, C., Hewitt, C.A., Huang, H., Montgomery, D.S., Xu, J., Carroll, D.L. (2015) *Phys. Chem. Chem. Phys.* **17**, 8591.

60 Yang, Y., Lin, Z.-H., Hou, T., Zhang, F., Wang, Z.L. (2012) *Nano Res.* **5**, 888.

7

Industrial Perspective on Printed Batteries

Patrick Rassek[1], Michael Wendler[2] and Martin Krebs[3]

[1] *Hochschule der Medien (HdM), Innovative Applications of the Printing Technologies (IAF/IAD), Stuttgart Media University, Germany*
[2] *ELMERIC GmbH, Rangendingen, Germany*
[3] *VARTA Microbattery GmbH, Innovative Projects, Ellwangen, Germany*

7.1   Introduction

The demand for flexible thin film batteries is increasing as new portable consumer electronics products, wearables as well as medical and pharmaceutical sensing devices, are developed, with this predicted to reach market maturity within the next few years [1–4]. To meet the requirements of this new category of flexible products, the batteries powering these devices also must be flexible. Conventional consumer batteries, which had been characterized and optimized over decades and thus are reliable power sources, are available in a variety of types, sizes, configurations and electrical specifications. Flat cells, cylindrical configurations and button configurations of batteries are predominant in the household and commercial sectors [5]. With respect to these new applications, the disadvantage of conventional batteries is their rigid casings. The integration of conventional batteries into flexible circuits requires on the one hand the use of discontinuous pick-and-place technologies, which contradicts a highly preferred cost-effective manufacturing process, while on the other hand, the use of conventional batteries in this type of product results in inflexibility of the products.

Using printing technologies for the production of thin film batteries offers many advantages compared to the conventional coating process [6]. Printing technologies are characterized by a large range of freedom with respect to the substrates and materials that can be processed. A wide range of polymer-based single-layered films or multilayered composite substrates with different

Printed Batteries: Materials, Technologies and Applications, First Edition.
Edited by Senentxu Lanceros-Méndez and Carlos Miguel Costa.
© 2018 John Wiley & Sons Ltd. Published 2018 by John Wiley & Sons Ltd.

properties as regards physical stability as well as gas and vapor permeation rates are available on the market. Furthermore, various primary and secondary electrochemical systems where the active masses of the electrodes are transferable into printing pastes can be processed by printing technologies. The possibility of producing batteries with customized layouts and defined operating voltages, capacities and service lives without extensive modification of assembly lines causing considerable conversion costs on coating machines makes printing technologies even more interesting as a manufacturing method. Process parameters evaluated in laboratory scale can be scaled up to industrial printing processes using roll-to-toll technologies. Therefore, production capacities of graphical printing machinery already installed can be used. The modular construction of many printing machines enables the installation of special equipment like pretreatment units, drying ovens or sealing units. This enables adaptation of printing technologies to the special requirements of battery manufacture. The ability to realize parallel and series arrangements of batteries by simply replacing printing forms and assuming verified process parameters allows a quick response to market demands or layout changes requested by the customer.

The printability of well-known and emergent electrochemical systems which can be processed in an ambient environment is currently the subject of research, with many different approaches and manufacturing techniques being investigated [6–12]. Still, many challenges remain until fully printed batteries can be produced in high quantities. This includes the optimization of printing pastes with respect to particle sizes and extended processing times as well as the investigation of appropriate sealing methods, physical and chemical stability of the printed functional layers, and a reliable in-line quality assurance system that detects and sorts out batteries containing errors.

This chapter provides a detailed overview and an evaluation of printing technologies that can be used to produce printed thin film batteries in high quantities. Advantages and disadvantages of particular printing technologies will be discussed, with consideration of the technological aspects, battery-specific requirements and output capacities. Furthermore, the current status of and recent progress in the fabrication of fully printed zinc-based batteries will be presented, specifying the obstacles that must be overcome for a laboratory-scale printing process to be transferred to an industrial printing process. Concepts of possible applications for printed batteries in flexible electronic devices will be discussed as well as the industrial perspective of these applications.

7.2 Printing Technologies for Functional Printing

The worldwide availability of printing technologies is driven by a steady demand for products like newspapers, magazines, books and packaging. Every printing technology has its own specific field of application due to

Table 7.1 Major quality aspects in the fields of graphical printing and functional printing.

Graphical printing	Functional printing
dot gain/dot loss	ink deposition and printed-layer thickness
optical density	line widening
line-edge sharpness	line-edge sharpness
register accuracy	register accuracy
distortion	distortion
layer homogeneity	electrical conductivity/resistivity
	surface roughness/porosity
	layer homogeneity

characteristic machine constructions and configurations, printing forms, ink rheology, type of ink transfer and throughput capacity. However, functional printing sets quality requirements for printing technologies that are different to those in graphical printing. This is why quality assurance for these functional layers is another challenge in the field of printed electronics. Beyond visual inspection, characterization methods with respect to electrical functionality of the printed patterns, lines or layers must be developed and installed in printing machines. Willmann *et al.* argue that the combination of "[...] all necessary manufacturing steps into one in-line production is the long-term goal of printed electronics" [13]. Achieving this goal requires a high degree of extensibility of printing machinery. Assuming mass-production of printed electronics applications with roll-to-roll machines, in-line quality assurance systems have to work reliably at high printing speeds, as error-containing products must be detected, marked and sorted. Table 7.1 lists major quality aspects that need to be considered in the fields of graphical and functional printing.

Additionally, a variety of products exist in the field of printed electronics with different requirements regarding print resolution, layer thickness, surface homogeneity and the types of substrate that can be processed. The choice of the preferred printing technology depends on product-specific properties and the materials that need to be processed.

For example, the front-side metallization of silicon solar cells requires the printing of fine line contact fingers with a high aspect ratio to reduce shading loss and enhance electrical conductivity at the same time. Contact fingers with line widths in the range of 30 μm, often realized with double prints to achieve sufficient electrical conductivity, can only be printed assuming an efficient registration control and customized printing pastes. Special printing equipment and highly viscous screen printing silver pastes with a solid content up to 90%

Table 7.2 Comparison of selected parameters of the common printing technologies. Adapted from [2], Table 1. Reproduced with permission of John Wiley and Sons, Ltd.

Printing technology	Print resolution [μm]	Print speed [m min^{-1}]	Wet-film thickness [μm]	Ink viscosity [mPa s]
Flexography	30–75	50–500	0.5–8	50–500
Gravure	20–75	20–1,000	0.1–5	50–200
Offset	20–50	15–1,000	0.5–2	20,000–100,000
Screen	30–100	10–100	3–300	500–70,000
Inkjet	20–50	1–100	0.3–20	1–40

and silver particles with sizes of less than 5 μm are available for those applications [14]. In contrast the electrodes of printed thin film batteries are realized by processing printing pastes containing highly-abrasive particles with sizes up to 75 μm. Whereas the resolution of the electrodes to be printed is insignificant in a certain range, layer thicknesses of 100 μm and more are needed to ensure suitable electrode capacities. Yet, sufficient registration control is needed in this field of application as stack-type batteries and coplanar batteries can be manufactured by multilayer printing. Table 7.2 gives a comparative overview of selected parameters of the traditional printing technologies.

Since Table 7.2 shows significant differences in the values of the ink viscosity, print resolution and wet-film thickness that can be achieved, a more detailed view of the characteristics of the individual printing technologies regarding the manufacture of batteries will be given. Since there is currently enormous progress in the development of digital printing solutions for functional printing, only printing technologies with permanent printing forms will be discussed.

7.2.1 Flexography

In flexographic printing a flexible rubber or photopolymer relief printing plate or sleeve is used as printing form. Printing plates are available in a variety of parameters with respect to plate thickness, plate hardness and relief depth. A specific plate type is selected according to the requirements of the layout to be printed as well as the surface properties of the substrate to be printed. The printing elements of the printing plates are raised and the relief is formed by photopolymerization. After mounting the flexible printing plate onto the plate cylinder, the raised parts of the printing plate are inked up with an anilox roller made of chrome or ceramics. The ink transfer can be controlled by selecting an anilox roller with a specific dip volume and screen frequencies of up to 450 lines/cm. A doctor blade removes excessive ink from the anilox roller before the printing plate is inked up with a reproducible amount of ink [15].

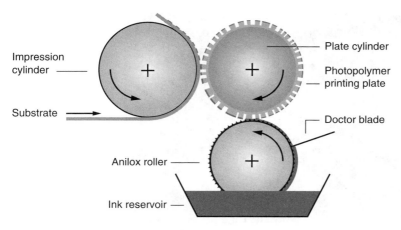

Figure 7.1 Construction principle of a flexographic printing unit.

The substrate to be printed runs on a hard impression cylinder while the plate cylinder with the soft printing plate being mounted is pressed against the substrate. Ink transfer onto the substrate is realized by ink-splitting from the raised parts of the printing plate to the substrate. Multicolor web-fed printing machines with 8–10 printing units and high throughput capacities with printing speeds of up to 500 m/min are predominant in the field of flexographic printing. Sheet-fed printing machines are also available and are used for special applications [16]. Figure 7.1 shows the main features of a flexographic printing unit.

Various types of substrate can be processed by flexographic printing including paper, cardboard, metal and plastic film, as well as composite substrates, as this printing technique is traditionally used to produce packaging. Additionally, the modular construction of the printing machines and the soft photopolymer printing plates that can realize line widths up to 30 µm make this printing technology attractive for printed electronic applications like the printing of contact fingers of silicon solar cells [16]. The polymer-based printing plates and the ceramic anilox roller enable production of batteries free of contamination by the active masses of the electrodes. Low viscosity conductive and non-conductive printing inks based on water or organic solvents have been developed and optimized for flexographic printing. Figure 7.2 shows an example of a flexographic printing plate with a detailed view of the raised printing elements.

Recent investigations demonstrated interest in using flexographic printing as a printing technique for zinc-based batteries [17, 18]. Printed wet-film thicknesses with a maximum of 8 µm can be achieved with a single printing stroke. To achieve printed-layer thickness of the electrodes necessary to ensure electrical performance, multiple printing steps must be performed on the same substrate.

Figure 7.2 Example of a nyloflex® digital photopolymer printing plate manufactured and distributed by Flint Group with detailed view of the raised printing elements. Reproduced with permission of Flint Group Germany GmbH (Stuttgart, Germany).

This requires either the installation of additional printing units printing the same layout or multiple printing cycles with precise registration control. For the production of fully printed batteries with printed current collectors, electrodes and electrolyte, printing machines with more than ten printing and drying units are needed; however, this is not an economically feasible manufacturing process for printed batteries.

7.2.2 Gravure Printing

The printing form of a rotogravure printing machine consists of a hard cylinder mainly made of a steel core coated with a copper layer acting as the image carrier. The printing elements are cells which are engraved into the copper surface of the cylinder by laser engraving, mechanical engraving or an etching process, while the non-printing elements remain on the original surface level of the cylinder. For long print runs the engraved cylinder can be chromium plated to extend run resistance. The printing unit of a rotogravure printing machine consists of the printing cylinder, an ink reservoir, an impression cylinder and a doctor blade. Before printing, the whole printing cylinder is flooded and the engraved cells are filled up with low viscosity ink. Before contact between printing cylinder and the substrate is made, a doctor blade removes excessive ink from the non-printing areas and ensures that ink remains only in the engraved cells. Ink transfer from the cells onto the substrate is realized by applying high printing pressure of approximately 3 MPa [15]. Sometimes ink transfer is optimized by use of an electrostatic printing assist system which

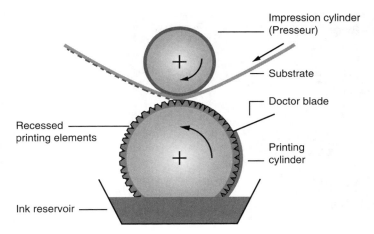

Impression cylinder (Presseur)

Substrate

Doctor blade

Recessed printing elements

Printing cylinder

Ink reservoir

Figure 7.3 Construction principle of a rotogravure printing unit.

supports the emptying of the cells [19]. Gravure printing is a high-speed print-ing technology which is mainly used for high-volume print runs of magazines and packaging, and decorative and securities printing. Predominantly, multi-color roll-to-roll printing machines are installed but sheet-fed machines also exist. The construction principle of a printing unit of a rotogravure printing machine is illustrated in Figure 7.3.

The advantages of rotogravure printing are the flexibility of the printing format and the potential to print products in large quantities. While printing cylinders are usually available in widths between 200 mm and 4,320 mm [19], the manufacturing costs of the printing forms are relatively high compared to other printing technologies, as a single printing cylinder is needed for every separation to be printed. Figure 7.4 shows a detailed view of a chromium-plated rotogravure printing cylinder with its recessed printing elements.

The possibility of simultaneously printing thin layers at high print resolution and at large scale intensified research into the use of gravure printing in the area of printed electronics in recent years. The fabrication of organic solar cells [20] and wireless sensor-signage tags [1] and the patterning of flexible sub-strates with silver nanowires [21] or graphene inks [22] is just a selection of gravure printing applications currently being investigated. The advantage of printing thin layers at high print resolution represents a disadvantage when batteries have to be printed. As discussed earlier (see section 7.2.1) electrodes of batteries require a certain layer thickness to achieve sufficient capacity. Furthermore, contamination of printing pastes consisting of the active masses of the electrodes cannot be ruled out as printing cylinders are made of steel, copper and chromium. This seems to be another reason why gravure printed batteries have not been reported as yet.

Figure 7.4 Detail view of a chromium-plated rotogravure printing cylinder with recessed printing elements.

7.2.3 Offset Printing

In offset printing the printing elements and the non-printing elements are virtually located on the same layer of the printing form. The adhesion of the high viscosity inks on the printing elements is caused by differences in the value of surface energy compared to the non-printing elements of the aluminum-based printing form. The imaging of the printing plate is usually realized by photolithography or in some cases by contact print copy with a film. In conventional offset printing, the printing unit consists of a dampening unit, a multi-roller inking unit, a plate cylinder with the printing form mounted, a blanket cylinder which transfers the ink from the printing plate onto the substrate, and an impression cylinder (see Figure 7.5). Before the ink is transferred onto the printing plate the whole plate is damped with a water-based dampening solution. The non-printing elements exhibit hydrophilic behavior, which results in ink-repellant properties, whereas the printing elements can accept the printing ink due to their oleophilic properties. Figure 7.6 exemplarily shows an offset printing form with printing and non-printing elements. During printing, the ink is initially transferred onto the blanket cylinder before being transferred onto the substrate by ink-splitting at a high printing pressure of about 1 MPa. Waterless offset printing technology is also available. Silicone-coated printing plates are imaged by photolithography to realize the distinction between ink-accepting and ink-repelling elements. Printing machines are available in sheet-fed and web-fed constructions [15, 23].

The multi-roller construction of offset printing units impedes the processing of functional printing pastes containing coarse particles of electrode materials. Another restriction of this printing technique in the desired field of application is

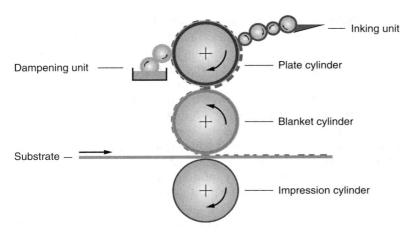

Inking unit

Dampening unit

Plate cylinder

Blanket cylinder

Substrate

Impression cylinder

Figure 7.5 Basic construction of a conventional offset printing unit.

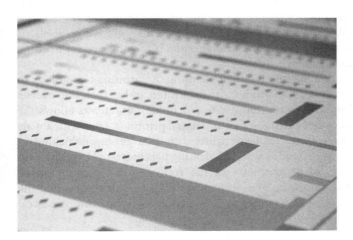

Figure 7.6 Detail view of an offset printing plate with the printing elements (green colored) virtually located on the same layer as the non-printing elements (white area).

the limitation of the layer thickness that can be achieved with a single printing stroke. Furthermore, the presence of the water-based dampening solution and the requirements regarding viscoelastic behavior of the printing pastes prevents offset printing technology being used as a manufacturing method for printed batteries.

7.2.4 Screen Printing

In screen printing the making of the printing form and the ink transfer onto the substrate is completely different as compared to the printing technologies described in the sections above. Woven meshes made of polyester, polyamide or stainless steel with defined counts of threads having defined diameters provide the basis for making a screen printing form. Initially, the meshes are stretched by applying a selected tension and then glued to a frame, made mainly of aluminum or steel, by means of adhesives. Printing elements are characterized as open mesh areas allowing the ink or paste to be transferred onto the substrate. The patterning of the printing form with a specific layout is realized by stencil making. Therefore, the open mesh area is coated uniformly with a photosensitive emulsion which can be exposed with UV irradiation by contact print copy after drying. As right reading positives are used as a film, the unexposed areas of the emulsion can be washed out after the exposure has finished, thus creating the printing elements of the stencil. The printing unit of a flatbed screen printing machine, as shown in Figure 7.7, consists of the printing form, a doctor blade and an angled squeegee made of polyurethane or combinations of rigid materials and polymers. Besides flatbed screen printing, which is predominant if printing jobs with precise register accuracy need to be done, a variety of machine configurations are available for special printing jobs, including for shaped objects like glassware, tools or barrels. Rotary screen printing is also available and is mainly used for printing ceramic tiles or labels [24].

Screen printing offers the possibility of adapting the printing technology to the specific requirements of the products to be printed, rather than the reverse. Understanding of the interaction of essential printing parameters allows the processing of a variety of substrates and functional printing pastes with a wide range of viscosity as well as multilayer prints with different layer thicknesses within one printing cycle. Modification of printing parameters like mesh count and thread diameter, blade geometry of the squeegee and printing speed enables control of the print resolution and the thickness of the printed ink layer, as well as the coarseness of the particles dissolved in the printing pastes.

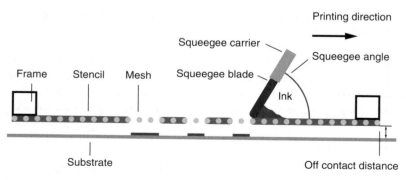

Figure 7.7 Main elements of a flatbed screen printing unit.

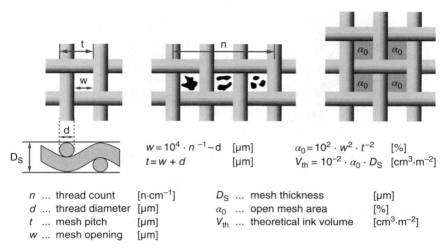

$$w = 10^4 \cdot n^{-1} - d \quad [\mu m]$$
$$t = w + d \quad [\mu m]$$

$$\alpha_0 = 10^2 \cdot w^2 \cdot t^{-2} \quad [\%]$$
$$V_{th} = 10^{-2} \cdot \alpha_0 \cdot D_S \quad [cm^3 \cdot m^{-2}]$$

n ... thread count	[n·cm^{-1}]	
d ... thread diameter	[µm]	
t ... mesh pitch	[µm]	
w ... mesh opening	[µm]	
D_S ... mesh thickness	[µm]	
α_0 ... open mesh area	[%]	
V_{th} ... theoretical ink volume	[cm^3·m^{-2}]	

Figure 7.8 Geometrical parameters of screen printing meshes.

A detailed view of mesh parameters enables a deeper understanding of the flexibility of screen printing technology with respect to the printing jobs and materials that can be processed. The mesh geometry is defined by the thread diameter d, the mesh count n and the mesh opening w, as illustrated in Figure 7.8. These parameters enable calculation of the open mesh area α_o representing the percentage of the ink-permeable area affecting ink-release characteristics and the thickness of the printed layers. Another essential factor is the theoretical ink volume V_{th}, which allows an approximate calculation of ink consumption and the maximum achievable thickness of the printed wet layer on non-absorbent substrates for each individual screen printing mesh.

The main factors that need to be considered when printed electronic devices are realized are the print resolution with respect to line-edge sharpness and the layer thickness of the deposited printing pastes on the substrate. The determination of mean particle sizes of printing pastes is necessary in advance for selection of an appropriate screen printing mesh capable of processing the particular printing paste. Sufficient ink penetration can be ensured by selecting meshes with a mesh opening that is three times the value of the mean particle size of the printing paste [24].

The resolution is an important criterion in the area of fine line printing of electronic circuits and when parallel battery architectures are realized. The dependency between the value of thread diameter and mesh opening determines the ability to print sharp edges. A screen printing mesh having a greater mesh opening than the thread diameter can be used to print fine lines

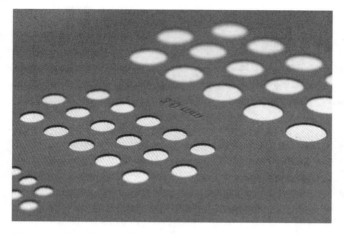

Figure 7.9 Detail view of a screen printing form with a thick film stencil (printing side). Printing elements are characterized as open mesh areas allowing the printing paste to be transferred onto the substrate.

at a higher resolution. Assuming a proper stencil technique, the characteristic visual appearance of screen-printed lines is minimized as is the risk of interruptions affecting electrical functionality. The availability of a variety of screen printing meshes made of dissimilar materials in various mesh counts and thread diameters allows ideal adjustment of the screen printing equipment to the desired layer thickness, line-edge sharpness and particle sizes of the preferred printing pastes. This freedom of process design is the major advantage of screen printing technology and the main reason for the widespread use of this technology for printing functional layers. Figure 7.9 shows a detail view of the printing side of a screen printing form made with a thick capillary film.

Characterized as a cost-effective and versatile manufacturing method, screen printing technology is now used on an industrial scale for realizing the frontside metallization of crystalline silicon solar cells [25], for antennas for automotive applications [26] and for sensors for medical applications [27]. It is also a promising technology for economically producing flexible batteries on a large scale, as the technical features of the printing forms allow an adjustment of the mesh geometry according to the desired printed film thickness and the mean particle size of the active masses of the electrodes. In recent years, research has focused on the development of screen printing pastes of well-known electrochemical systems for flexible thin film batteries for wearables or other portable devices [9, 11]. As fully screen-printed zinc-carbon batteries and nickel-metal hydride (NiMH) batteries have been successfully demonstrated [6, 28], other electrochemical systems like primary or secondary zinc-air batteries are currently drawing the attention of several research groups [7, 12], since these batteries consist mainly of eco-friendly materials.

7.2.5 Conclusion

As can be seen, substantial technological differences with respect to ink transfer, ink-splitting, print resolution and materials accessible persist between the traditional printing technologies using a permanent printing form. Existing variations in the construction principles of the printing forms, printing units and printing machines determine individual process characteristics affecting quality of registration control, layer thickness achievable with a single printing stroke, and possible risk of contamination of the printing pastes of the electrodes.

The flexibility and extensibility of screen printing technology meet the requirements for the manufacture of printed batteries. The characteristics of the printing forms allow individual control of printed-layer thickness and print resolution, even within one printing cycle. Many functional printing pastes have been optimized for screen printing technology and variation of the mesh geometry of the printing form enables formulation of printing pastes based on different types of solvents and particle sizes. The extension of screen printing machines with specially modified equipment from other technologies in the field of packaging allows the integration of various in-line sealing methods. Compared with other printing technologies, screen printing has the highest level of vertical integration, assuming fully printed batteries on an industrial scale, and is therefore the technology of choice in this field of application.

7.3 Comparison of Conventional Battery Manufacturing Methods with Screen Printing Technology

Production and assembly lines for the manufacture of conventional batteries with standardized layouts consist of multiple stations of highly innovative and productive machine components. Only a few specialized engineering companies are able to construct technologically advanced production and assembly lines with output capacities of 100 or more batteries per minute. All processing steps needed to produce complete batteries are included within the production line, as well as in-line quality monitoring systems. Figure 7.10 exemplarily illustrates a fully automated production line for lithium-ion batteries, indicating the vital process steps. Both electrodes are realized by a continuous coating process (e.g. slot die coating) of the active masses on carrier substrates, while current collectors are integrated by pick-and-place technologies of pre-punched metal

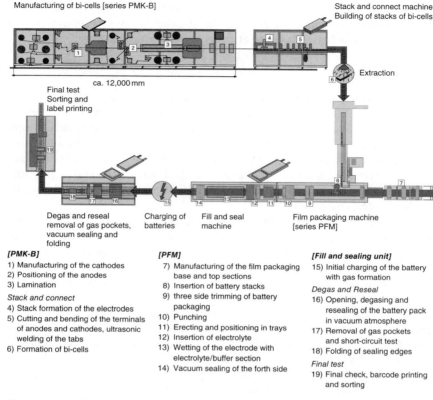

Manufacturing of bi-cells [series PMK-B]

Stack and connect machine
Building of stacks of bi-cells

Extraction

ca. 12,000 mm

Final test
Sorting and
label printing

Degas and reseal
removal of gas pockets,
vacuum sealing and
folding

Charging of
batteries

Fill and seal
machine

Film packaging machine
[series PFM]

[PMK-B]
1) Manufacturing of the cathodes
2) Positioning of the anodes
3) Lamination

Stack and connect
4) Stack formation of the electrodes
5) Cutting and bending of the terminals
   of anodes and cathodes, ultrasonic
   welding of the tabs
6) Formation of bi-cells

[PFM]
7) Manufacturing of the film packaging
   base and top sections
8) Insertion of battery stacks
9) three side trimming of battery
   packaging
10) Punching
11) Erecting and positioning in trays
12) Insertion of electrolyte
13) Wetting of the electrode with
    electrolyte/buffer section
14) Vacuum sealing of the forth side

[Fill and sealing unit]
15) Initial charging of the battery
    with gas formation
Degas and Reseal
16) Opening, degasing and
    resealing of the battery pack
    in vacuum atmosphere
17) Removal of gas pockets
    and short-circuit test
18) Folding of sealing edges
Final test
19) Final check, barcode printing
    and sorting

Figure 7.10 Fully automated production line for the production and assembly of lithium-ion batteries. Reproduced with permission of Harro Hoefliger Verpackungsmaschinen GmbH (Allmersbach im Tal, Germany). (*See insert for color representation of the figure.*)

foils. Staples of cells are formed and packed into pouches before the electrolyte is inserted in a precise quantity into the nitrogen atmosphere by dispensing technology. The final steps include preliminary sealing of the pouches followed by degassing and resealing of the pouches before functional checks and initial formation of the batteries are performed.

As shown in Figure 7.10, the dimensions of production lines for the manufacture of lithium-ion batteries require significant investment costs with respect to machine configurations and corresponding infrastructure. The capacities of these machines are sufficient for realizing high quantities of standardized batteries at low reject rates. These batteries are needed to meet the growing demand for energy storage systems for smartphones, tablets, notebooks and many other portable electronic devices. Switching to other battery designs or materials like carrier substrates causes considerable conversion costs for fully automated production lines, with an initial evaluation of each individual process parameter.

Compared to this technology, investment costs for sheet-fed or rotary screen printing lines with multiple printing stations are modest. While screen printing technology offers the possibility of switching battery designs in a short time at low cost, all process steps shown in Figure 7.10 must be transferred into printing steps. For example, the positioning of pre-punched metal foils for the realization of current collectors can be replaced by an initial printing of highly conductive silver pastes before the electrodes are deposited at the subsequent printing stations. Sealing procedures can be substituted by printing adhesives, lamination or ultrasonic welding processes that are known from the area of packaging technology. The modular construction of screen printing machinery offers a high degree of extensibility which allows integration of a variety of process technologies into the manufacturing line. Figure 7.11 gives an overview of a possible configuration of a manufacturing line for the production of zinc-based batteries consisting of two roll-to-roll screen printing machines having multiple printing units, followed by drying units and a sealing unit representing the final assembly steps for the batteries.

As can be seen, there are significant differences in the two technologies with respect to the level of vertical integration, output capacity, flexibility of processible materials and in-line quality monitoring and assurance. Competition between the two manufacturing methods can be excluded as different fields of

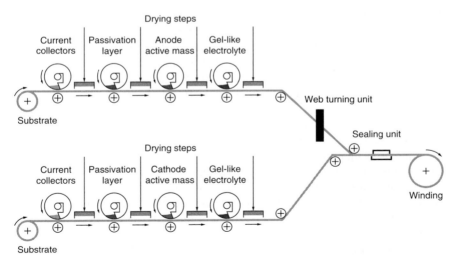

Figure 7.11 Exemplary presentation of two rotary screen printing lines and a sealing unit for the manufacture of zinc-based batteries in stack configuration. (*See insert for color representation of the figure.*)

application are served by the respective technologies. Screen printing is a promising technology when flexible thin film batteries with changing layouts must be produced at low costs. Therefore, further research work with respect to the adaptation of printing technology and materials to the special requirements of battery manufacture is a necessity.

7.4 Industrial Aspects of Screen-printed Thin Film Batteries

7.4.1 Layout Considerations

The transformation of conventional consumer battery architectures with rigid casings to printed battery layouts requires the development of new construction principles for batteries. Printed batteries are predominantly generated in layers beginning with the application of current collectors. Two architectures, sandwich and parallel, have been proven at laboratory scale and are suitable for industrial-scale production of fully printed batteries. At the production level, differences in the two architectures exist in the requirements regarding the standard of quality of printing machinery and equipment. The electrical performance characteristics of both construction methods in relation to short-circuit currents, extractable capacities and shelf-lives is still the subject of research.

7.4.1.1 Sandwich Architecture (Stack Configuration)

All functional elements of printed batteries having a sandwich architecture are successively generated in printing layers either on the same sheet of the carrier substrate or on one individual carrier substrate for each electrode, with both substrates brought together at the sealing unit, representing the final process step and thus the encapsulation of the battery (see Figure 7.11). Figure 7.12 depicts a schematic drawing of the construction method for a printed sandwich-type battery naming all required components. If printed on the same layer of a sheet, which is the preferred manufacturing method for this architecture, one printing unit followed by a drying unit is needed for every functional layer printed. Folding techniques must be applied after the separator is either printed or placed by pick-and-place technology on at least one electrode, ensuring contact of both electrodes with the separator and the ionic conductive electrolyte. Sealing and encapsulation of the batteries can be realized by a patterned printing of adhesives.

The arrangement of the terminals on the same layer facilitates the electric connection of the battery with other printed or non-printed electronic devices. The option of printing sandwich-type batteries sets lower requirements for the registration control of the printing machines and allows

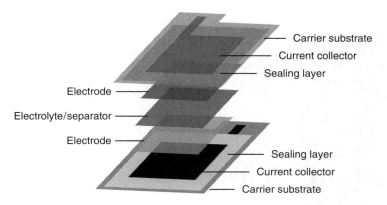

Carrier substrate
Current collector
Sealing layer

Electrode ———

Electrolyte/separator ———

Electrode ———

Sealing layer
Current collector
Carrier substrate

Figure 7.12 Schematic drawing of the layered construction principle of a printed sandwich-type battery in stack configuration (single cell).

manufacture with both sheet-fed and roll-to-roll screen printing machinery with high throughput capacities. It also represents a reduced risk of contamination with particles of the opposite electrode since the distance between the electrodes on the printing substrate can be significantly larger than that in batteries with a parallel architecture. The realization of this battery architecture by screen printing technology allows production of single-cell and multi-cell configurations of batteries in series or parallel arrangements exhibiting specific voltages and capacities.

Figure 7.13 shows screen-printed electrodes of an NiMH battery before the separator and electrolyte are placed. Current collectors consisting of printed silver and additional carbon black passivation layers have been realized, to avoid any negative interaction of the silver particles with the electrochemistry of the battery.

7.4.1.2 Parallel Architecture (Coplanar Configuration)

Screen printing of batteries having a parallel architecture, also called coplanar architecture, sets higher requirements for the registration control of printing machines as well as for the quality standards of the printing forms. Sheet-fed screen printing machines with three-point alignments are preferred when coplanar batteries need to be printed since a more precise registration control can be assured. All functional layers, including both electrodes, are printed on the same layer of the sheet and separated only by a gap with a defined width. Electrical contact between any functional layers due to insufficient registration control has to be avoided as this would cause short-circuits resulting in malfunction of the battery. Contacts of the positive and negative terminals are located on the same layer, which facilitates the connection of the battery to other electronic devices.

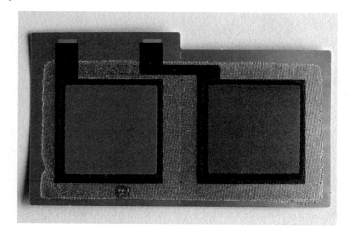

Figure 7.13 Printed electrodes of an NiMH battery (single cell) in stack configuration. The battery is finished after the separator is placed on one electrode followed by defined insertion of the electrolyte, the folding procedure and contact heat sealing of the printed adhesive layer (structured framing).

The main benefit of this battery architecture is the reduction in the use of printing forms, required printing units and subsequently drying units, which enables faster and more cost-effective production of these batteries on existing printing machinery without large modification. The printing of current collectors and passivation layers requires only one printing form, which also accelerates the drying procedures for the printed layers. Figure 7.14 gives an overview of the layers that must be printed during the manufacturing process of a printed coplanar battery.

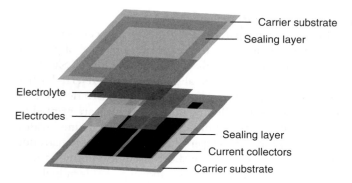

Figure 7.14 Schematic drawing of the layered construction principle of a printed coplanar battery with parallel architecture (single cell).

As folding processes are not needed for the encapsulation of coplanar batteries, possible integration of circuits of other electronic devices in the manufacturing process is substantially easier. For example, antennas for near-field communication (NFC) or RFID communication can be printed on the same carrier substrate with the particular layout being placed on the same printing form as is used for the current collectors, assuming the use of identical conductive inks for both structures. Another advantage of coplanar batteries is the absence of a separator, which simplifies the manufacturing process. The electrolyte can also be deposited by screen printing, whereby the amount of electrolyte being deposited can be controlled by selecting screen printing meshes with an adjusted theoretical ink volume.

As can be seen, printed coplanar batteries have a higher manufacturing depth compared to sandwich-type batteries, as all process steps can be realized by in-line screen printing technology followed by application of well-known packaging technologies (e.g. lamination process), which makes the printing of adhesives as sealants redundant. A higher risk of contamination of the pre-printed electrode areas with active masses of the opposite electrode printed at the following printing unit can be excluded when high-quality printing forms are used. Electrode sizes can be adjusted to the available printing format of the printing machines and desired battery capacity. Recently, fully printed single-cell and multicell zinc-carbon batteries in series and parallel arrangements have been demonstrated successfully. They are used to power autonomous working temperature sensors for a recording period of several months (see section 7.5.1). Figure 7.15 shows an example of a fully printed zinc-carbon battery (single-cell, ZnCl electrolyte) in an individually designed coplanar configuration and the corresponding discharge curve.

7.4.2 Carrier Substrates and Multifunctional Substrates for Printed Batteries

Before the most appropriate carrier substrate for the manufacture of printed batteries is identified and selected, consideration must be made of the desired electrochemical system, battery layout, and manufacturing methods with respect to printing machines, drying procedures and sealing technologies. Various types of substrate can be used for the production of printed batteries. This includes paper substrates, plastic films, metal foils and any combination of these materials resulting in multilayered composite substrates with defined properties of chemical and physical resistance as well as defined transfer rates of gases and water vapor from the ambient environment. Multilayered composite substrates like packaging pouches from the area of food packaging meet the demanding requirements that are set by printed batteries. Besides their flexibility, those substrates can be sealed and have adequate barrier capabilities as regards gases, moisture and various aqueous electrolyte systems. Despite

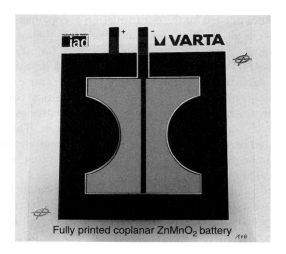

Fully printed coplanar $ZnMnO_2$ battery

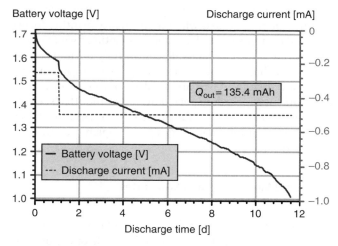

Figure 7.15 Fully printed coplanar zinc-carbon battery (single-cell, ZnCl electrolyte) with an overall thickness of 0.5 mm and the corresponding discharge curve. A capacity of 135.4 mAh was determined during discharge from the open-circuit voltage (OCV) to the cut-off voltage of 1.0 V. (*See insert for color representation of the figure.*)

those advantages and the high availability of these substrates in industrial quantities, most packaging materials have an asymmetrical construction of different types of plastic film on the interior and exterior sides, which results in curling behavior when necessary thermal treatment procedures are applied. To ensure proper alignment at all printing units, a symmetrical construction of suitable substrates is highly preferred in the area of printed batteries, especially when sheet-fed printing machines are used. Recently, various approaches have been made to develop specially designed substrates for the production of printed batteries.

7.4.2.1 Barrier Requirements and Material Selection

The chemical properties of the particular electrochemical system to be printed sets the requirements regarding the selection of materials for the manufacture of a proper substrate. Excellent barrier capabilities are required to avoid transfer of various gases and water vapor into and out of the encapsulated electrochemical systems of many batteries.

For example, the electrochemical systems of NiMH batteries and alkaline zinc-carbon batteries are sensitive to gas and water vapor transfer into and out of the battery. On the one hand, those batteries must be protected against the penetration of carbon dioxide (CO_2) from ambient air into the battery as this would cause and increase oxidation of the aqueous potassium hydroxide (KOH) electrolyte to crystalline potassium carbonate (K_2CO_3), resulting in a decrease of the electrical performance due to the insulating behavior of the carbonates. On the other hand, evaporation and loss of water from the aqueous electrolyte forces a drying out of the battery with a decrease in the ionic conductivity of the electrolyte. In contrast to the electrochemical systems of those batteries, the manufacture of zinc-carbon batteries with aqueous or gel-like zinc chloride as an electrolyte is less demanding regarding the encapsulation specifications of the substrates to be printed. As zinc chloride shows hygroscopic behavior, it can accept water vapor from the ambient air in a certain range, which extends battery shelf-life and service life. Obviously, loss of the electrolyte due to insufficient sealing has to be avoided as well, as this would cause malfunction of the battery. Multilayered composite substrates made of thin aluminum foils enclosed by protecting plastic films meet the requirements for impermeable carrier substrates for printed alkaline batteries. While the aluminum layer successfully impedes the transfer of gases and water vapor into the battery, the surrounding plastic films protect the aluminum layer from interaction with the electrolyte and enables sufficient printability at the same time due to a better wetting behavior for printing inks. In the area of zinc-carbon batteries with zinc chloride as an electrolyte, plastic films made of PE or PET offer sufficient encapsulation behavior to protect the electrochemistry from contamination and loss of electrolyte.

Electrochemical systems having an open cell design such as zinc-air batteries and other printable metal-air systems need a special combination of individual carrier substrates for each electrode. In the example of zinc-air batteries, the zinc anode should be printed on high-barrier substrates whereas the catalytic cathode needs oxygen from ambient air to maintain the electrochemical redox reaction that converts zinc to zinc-oxide during discharge. Therefore, porous membranes that are permeable to gases and simultaneously exhibit hydrophobic behavior preventing leakage of electrolyte are required. In this case, the presence of CO_2 in ambient air is another challenge as it supports the degradation and carbonation of the electrolyte consequently clogging the pores of the porous cathode with potassium carbonate, which results in a decreased cycle stability of this electrochemical system. While porous membranes made of PTFE are predominantly used as functional carrier substrates for cathode manufacture in conventional zinc-air batteries, the printability of these membranes represents a major challenge. The realization of printed primary and secondary zinc-air batteries with alternative substrates to PTFE in the manufacture of gas diffusion cathodes is currently the subject of research by several research groups and seems to be the key factor in making this electrochemical system accessible by screen printing technology. Table 7.3 gives a comparative overview of printable primary and secondary electrochemical systems and their respective specifications for the construction of carrier substrates.

As can be seen from Table 7.3, the characteristics of the particular electrochemical system principally define the specifications of suitable materials for utilization as carrier substrate. Besides these material standpoints, additional factors that need to be considered for industrial production of printed batteries are the process stability of the respective substrates during printing and the subsequent application of drying procedures. Those standpoints will be discussed in the upcoming section.

7.4.2.2 Process Requirements of Qualified Materials

The thickness of the carrier substrate mainly defines the overall thickness of the manufactured battery and also affects its flexibility. Assuming low-cost production of flexible batteries having high volumetric energy densities, thin substrates are most likely to be preferred, but the deficit of suitable barrier capabilities contradicts their usage. Thin substrates like single-layered plastic films also complicate the processing of the highly viscous screen printing inks that are predominantly used in the area of printed batteries. Thin substrates tend to stick at the printing side of the screen during ink release. Registration control with proper alignment of the substrate at the vacuum table of sheet-fed printing machines is also negatively affected due to insufficient handling capability. Furthermore, a certain layer thickness of heat-sealable plastic is required when hot-sealing techniques or ultrasonic welding processes from the area of packaging technology are applied for encapsulation. Therefore, multilayered composite substrates are the material of choice in the field of printed batteries. Figure 7.16 depicts a schematic diagram of a multilayered and paper-based composite substrate exclusively developed and manufactured by Schoeller Technocell (Osnabrück, Germany) for the production of printed NiMH and zinc-carbon batteries.

Table 7.3 Presentation of selected electrochemical systems that can be realized by screen printing technologies at ambient environment and their specifications for carrier substrates.

Electrochemical system	Battery system	Electrolyte	Demanded barrier capabilities	Required type of carrier substrate
Nickel-metal hydride	Secondary	Aqueous or gel-like KOH, alkaline	CO_2, electrolyte, water vapor	Multilayered composite substrates made of plastic foils and metal foils
Zinc-carbon	Primary	Aqueous or gel-like KOH, alkaline, Leclanché	CO_2, electrolyte, water vapor	
Zinc-carbon	Primary	Aqueous or gel-like ZnCl, zinc chloride	electrolyte	PET, PE
Zinc-air	Primary	Aqueous or gel-like KOH, alkaline	CO_2, electrolyte, water vapor	Zinc anode: multilayered composite substrates made of plastic foils and metal foils Air-cathode:
Zinc-air	Secondary	Aqueous or gel-like KOH, alkaline and other electrolytes	CO_2, electrolyte, water vapor	Porous but hydrophobic membrane

35 GSM LDPE ——

12 µm ALU FOIL ——

35 GSM LDPE ——

63 GSM PAPER ——

35 GSM LDPE ——

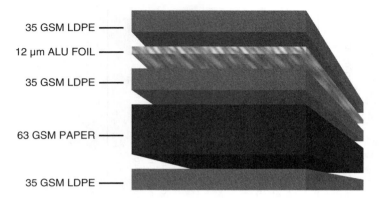

Figure 7.16 Construction of a multilayered and paper-based composite substrate (p_e: smart paper type 4) developed and manufactured by Schoeller Technocell GmbH & Co. KG. GSM: Gram per square meter. Reproduced with permission of Schoeller Technocell GmbH & Co. KG (Osnabrück, Germany).

The combination of dissimilar types of materials in a composite substrate as depicted in Figure 7.16 results in a typical layer thickness of 0.2 mm or more with an extended stiffness of the substrate compared to single-layered films. Whereas this thickness value seems quite high, neither flexibility nor the application of folding techniques are restricted, enabling the manufacture of batteries having sandwich or parallel architectures. The major benefit of this construction is represented by a reinforcement of the composite substrate with the enclosed paper layer, which provides dimensional integrity when thermal treatment procedures are applied before and after printing steps. Initially, these thermal steps are necessary before printing to remove the internal stress of the materials and to avoid shrinkage of the substrate after the first printing unit, which would make proper alignment at subsequent printing units impossible. After initial thermal treatment, curling behavior of many composite substrates with asymmetric construction can be observed. The additional paper layer prevents the substrate from curling and rolling up. Further thermal treatment steps are required for heat curing of the functional printing inks establishing the percolative conductive network of the current collectors, passivation layers and electrodes.

The aluminum layer provides sufficient barrier capabilities against gases and moisture whereas the thickness of the outer plastic layers enables printability as well as the possibility of applying encapsulation techniques like heat sealing or ultrasonic welding processes without physical damage to the lower layers including the aluminum barrier layer. To enhance the wetting ability of the plastic surface to be printed, pretreatment methods like atmospheric plasma or corona that are conventionally used in graphical printing can be integrated in printing machines and applied before the first layer is printed. Another essential factor that needs to be considered is the safety of the battery regarding physical damage by possible leakage of the electrolyte during its life-cycle. While the construction of the composite substrate is able to withstand mechanical damage in a certain way, the sealing layer seems to be the weak point of printed batteries. Suitable encapsulation techniques that can be integrated with or connected to printing processes will be discussed in detail in section 7.4.6.

In conclusion, for printed batteries, many material aspects and the final application of the battery must be considered before the proper substrate can be selected and evaluated. The availability of a wide range of substrates is due to their widespread use in the field of food packaging. Modifications of existing carrier substrates as well as the development of new substrates can be realized in coordination with manufacturers of packaging material to achieve an optimum trade-off between economic aspects, barrier capabilities and process stability.

7.4.3 Current Collectors

Current collectors of conventional consumer batteries like lithium-ion batteries consist of pre-punched or tailor-made metal foils of aluminum and copper with the active masses of the respective electrode directly coated onto the surface of the foils. The transformation of those materials to the area of printed batteries requires the printing of electrically conductive but also electrochemically inert patterns. Printed electronic inks like silver inks or carbon inks have been optimized over years for screen printing applications and thus are promising candidates for the realization of printed current collectors.

After thermal drying procedures are applied, the sheet resistance of the printed layers should be in a low-single-digit ohmic range whereas the wetting ability of the surface layer needs to be sufficient at the same time to promote good adhesion for printing of electrode layers free of defects. This ensures low contact resistance between the active masses of the electrodes and the current collectors. Whereas the printing of silver inks represents an enormous expense factor with material costs of 1,000 €/Kg or more, other printing inks based on nickel or semi-conductive polymers like PEDOT:PSS exhibit either toxicity or low conductivity which impedes printing of those materials. The consumption of silver printing inks can be controlled and reduced to a minimum by selecting screen printing meshes with adjusted theoretical ink volumes. Another approach in reducing ink consumption is variation of the layouts of the current collectors. Whereas solid-tone patches in current collectors are highly ink consuming, the printing of grid-like structures may be a useful option when they are subsequently coated with carbon black ink, ensuring accessibility of electrons to the complete electrode layer. In any case, passivation of the silver layer is necessary to prevent chemical interaction of silver particles with the electrochemistry of the battery. Therefore, protection of the silver ink layers by extensive overprinting of carbon black ink with sufficient layer thickness has to be applied. Since carbon black exhibits excellent chemical resistivity against acidic and alkaline solutions, it additionally protects the silver layer from corrosion, which is of particular importance in the area of alkaline battery systems. Thermal treatment procedures must be adapted to the requirements demanded by the printing inks for heat curing as well as to thermal stability specifications of the carrier substrate. While most conductive printing inks need a curing time of 10–15 minutes at temperatures of 90–120 °C, printing speeds must be adjusted to the length of drying ovens to achieve adequate curing and electrical conductivity. Recent developments in the area of functional printing inks aim to reduce curing temperatures to values of 40–50 °C at shorter drying periods. This represents an enormous benefit regarding the materials available to printed battery manufacturing, as well as accelerating drying procedures through installation of shorter drying ovens, and also reduces energy costs.

To ensure long-term stability of printed current collectors resulting in extended service life for the manufactured batteries, the chemical and physical stability of the printed layers must be characterized individually in laboratory tests before larger printing runs are performed. The physical adhesion of printed current collectors and passivation layers to the particular carrier substrate can be examined by cross-hatch cutting or evaluation of folding processes, allowing visual quality control of the printed layers after the application of physical stress. Cyclic voltammetry allows the simulation of charge and discharge cycles in close succession enabling determination of the achieved chemical resistance of the printed passivation layers. Electrolytic corrosion of current collectors is mainly responsible for the delamination of those structures from the carrier substrate despite sufficient wetting behavior of the surface layer. Figure 7.17 shows electrolytic corrosion of the current collectors of a printed NiMH battery with subsequent delamination of the terminals from the carrier substrate. Further investigation and characterization of materials for the manufacture of current collectors for printed batteries exhibiting long-term stability represents an enormous challenge. While printability and high availability of the desired materials has to be ensured for cost-effective large-scale production, interaction of any component forming part of the construction of the battery with its particular electrochemistry must be excluded.

7.4.4 Electrodes

The main challenge of printed batteries is the formulation of printing inks based on the active masses of the electrochemical systems. Those raw materials are available in a variety of specifications with respect to particle size and

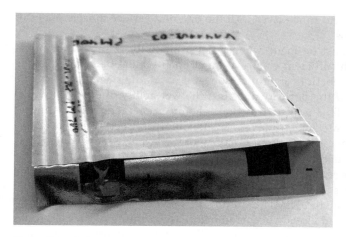

Figure 7.17 Delamination and electrolytic corrosion of the terminals of a printed NiMH battery with alkaline electrolyte (KOH).

particle form. The properties range from finely dispersed metal powders or oxides in the single-digit micron range to coarse particles of around $70-100\,\mu m$ having dissimilar surface morphologies. Spherical particles or flakes of different diameters are the predominant forms in the area of metals for the preparation of printing inks for negative electrodes while fine powders of oxides or alloys are used to prepare positive electrode printing inks. Figure 7.18 is a detail

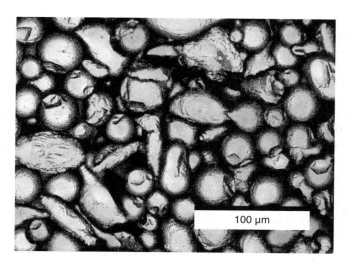

100 μm

Figure 7.18 Confocal laser scanning microscope image of a printed zinc anode layer with the different surface morphologies of the zinc particles clearly visible.

view of a printed zinc anode layer with the different surface morphologies of the zinc active mass.

Compromises must be made when good printability at satisfactory processing times with sufficient electrical functionality must be ensured at the same time. The selection of proper solvents or binding agents has to be adjusted in view of the porosity and thus the wettability of each individual active mass as well as adjusted to their chemical characteristics. Another requirement of printed electrode layers is flexibility in a certain range allowing for further processing without breaking of the layers, which would result in decreased electrical performance or malfunction of the batteries. For example, Wendler *et al.* proved that the formulation of printing pastes based on organic solvents results in reduced electrical performance parameters of printed zinc-carbon and NiMH batteries. It was observed that oxidation of the active masses had already occurred before the printing of the electrodes started. This chemical reaction was identified by determination of the electrical performance of the manufactured batteries and also by visual inspection of the printed electrode layers. After the ink formulation of solvents and binding agents was switched to water-based systems, a significant increase in the open-circuit voltages of the manufactured batteries compared to the theoretical values was observed. The results of comparative impedance measurements on batteries with electrodes realized with printing inks based on different types of solvents exhibited a decline in ohmic resistance in charged states when water-based printing inks were used. Those measurements represented metrological proof of the optimization of the electrochemical performance without quantifiable oxidation of the printing pastes prior to printing [28].

Another aspect that is often overlooked in the field of printed batteries is the adjustment of screen printing equipment to the special requirements of printing inks based on active masses for electrodes. Metallic contamination of the printing pastes has to be avoided in any case to ensure proper functionality of the electrochemical systems without parasitic side reactions negatively affecting the performance of the battery. This adjustment involves nearly all process steps from mixing of the components of the printing pastes, to stencil making and up to the printing of the electrodes. In the field of graphical printing, inks are often stirred up using dissolvers with metal discs to ensure proper homogenization of the inks with the pigments dispersed uniformly within the printing ink. To exclude any metallic contamination, metal discs need to be replaced by plastic discs or plastic-coated metal discs when electrode printing pastes are stirred. Printing pastes need to be stored in plastic containers and should be stirred up with plastic ink spatulas directly before printing. While doctor blades made of stainless steel seem to be noncritical, plastic alternatives should be considered as well to exclude any incalculable variations of electrical performance of the batteries within a patch.

The production of printing forms for printing electrodes requires specially adapted materials as well. Before printing forms are made, the geometry of screen printing meshes has to be adjusted to the particle sizes of the particular printing inks and to the desired electrode layer thickness that needs to be achieved. Furthermore, the type of solvent the printing paste is based on determines the type of emulsion or capillary film that can be used for stencil making. Assuming water-based printing inks, water-resistant emulsions with adequate physical stability against the coarse and abrasive particles of the active masses dispersed within the printing pastes need to be selected and validated with respect to desired stability and printing times. If this is ignored, physical stress caused by the movement of the printing ink and its dispersed abrasive particles by the squeegee and the doctor blade during the printing process can lead to a precocious destruction of the stencil with risk of contamination of non-printing areas with active masses. Reduced print quality with respect to line-edge sharpness, which is a requirement especially when coplanar batteries are printed, is also a result of a spoiled printing form.

The selection of printing meshes with defined mesh geometry in combination with the geometry of the installed printing squeegee affects and enables control of the printed-layer thickness of the electrodes. Depending on the particular mesh geometry, the viscosity of the printing paste and the geometry of the squeegee, uniform printed electrode layer thicknesses in the range of 60–100 µm have been achieved with single printing strokes. The challenge in transferring a laboratory-scale screen printing process to an industrial-scale printing process is represented by the deficit of adequate in-line quality assurance systems. Besides a visual inspection of the printed electrode layers with respect to surface defects, quality specifications should include continuous measurements of layer thickness and sheet resistance. While in-line monitoring systems exist in the field of graphical printing, ensuring identical reproduction of tonal values, these systems could also be used to recognize and mark electrodes having defects in surface homogeneity. Another major problem is represented by the deficit of working in-line measurement techniques with respect to printed-layer thickness, which is important to exclude any variations of capacities of the manufactured batteries within a patch. Irregular layer thickness across the printed layer is caused by a continuous evaporation of solvents out of the printing pastes, which affects rheological behavior and results in an increase in the ink viscosity. Progressive drying of the printing ink in the meshes of the printing form also has a negative effect on the quality of the printed electrode layers. Selective measurements that are performed manually to determine layer thicknesses are time consuming, but are still the method of choice in the field of quality assurance of printed batteries.

These problems are exacerbated by the need for consecutive manufacturing with subsequent assembly of the printed batteries, since printed electrode layers oxidize at an ambient environment if not stored in a controlled environment. Therefore, batches of primary batteries, which have the advantage of being fabricated in a charged state and thus are ready to use, should only be printed on demand as self-discharge cannot be avoided over storage time, resulting in a decreased OCV on the first day of usage. Secondary systems need an initial formation process with a charging procedure before the batteries can be used as a power source. As self-discharge of these systems before usage can be excluded, they are also susceptible to long storage periods as the electrolyte can support corrosive reactions inside the battery. Another factor that has significant impact on the shelf-life of printed batteries is proper encapsulation of the electrochemistry. In section 7.4.6, proven sealing methods will be discussed in detail in relation to industrial-scale production of printed batteries.

7.4.5 Electrolytes and Separator

The purpose of the electrolyte is to conduct the ions from one electrode to the other depending on their charge. On the other hand, an electrolyte should not conduct electrons, which would cause an internal short-circuit leading to a high self-discharge rate. For that purpose, the presence of ions and a certain ion mobility within the electrolyte is required. Usually the electrolyte is liquid with a conducting salt dissolved in an aqueous or non-aqueous solution at high concentration. In this case the ion conductivity is quite high. If the gap between the two electrodes is not guaranteed by mechanical means, this may lead to a short-circuit and a separator is needed. Usually the separator is made of fleece, woven or non-woven, and having high porosity. For special types of batteries like lithium-ion batteries, microporous membranes made of PE, PP and other plastic materials are used as separators.

It is possible to transform the electrolyte into a near solid state by adding binding agents to the extent that the electrolyte exhibits a rubber-like consistency. In this case the electrolyte can also be printed as a paste when coplanar batteries with parallel architecture are being printed. During the drying procedure the electrolyte nearly solidifies but retains its high ionic conductivity making the existence of a conventional separator redundant. The printing of electrolytes for the manufacture of sandwich-type batteries in stack configuration requires the addition of further additives acting as spacers to assure mechanical distance between the two electrodes. Those types of electrolyte/separator combinations can also be realized with screen printing technology but require additional modification of the printing equipment ensuring printability as well as homogeneous distribution of the coarse particles used to realize the mechanical distance within the printed layers.

Most recently, the investigation of solid electrolytes was reported. In this case the electrolyte forms a polymer layer with the ions moving along the polymer chain in the absence of any liquid phase. As yet, the resulting ion conductivity of these systems is far too low in the ambient temperature range, allowing use of these types of electrolytes only at elevated temperatures in all-solid-state batteries (ASSB) [29].

7.4.6 Encapsulation Technologies

Several sealing techniques mainly from the area of food packaging are available for the realization of encapsulations of printed batteries. Well-known technologies like contact heat sealing or ultrasonic welding can be integrated in production lines by connecting them after the final printing unit. To avoid interruption to or deceleration of the printing process, these sealing technologies have to perform at identical process speeds to those of the printing process. Additionally, suitable sealing methods must be selected according to their compliance with the carrier substrate and the particular thickness of the heat-seal polymer layer, as well as the sensitivity of the particular electrochemical system to mechanical stress and heat input. Another promising approach to the encapsulation of printed batteries is the structured printing of adhesives as sealant layers.

7.4.6.1 Screen Printing of Adhesives

Screen printing of adhesives is the preferred sealing method as it can be directly integrated into the existing manufacturing process allowing in-line manufacturing of printed batteries without large modification of the printing process and the printing machine. Another benefit is the variability in printing layouts by the replacement of the particular printing forms. The widespread use of screen-printable adhesives is typically unknown but the variety of usage ranges from self-adhesive products like decals to innovative applications in the automotive and electronics industries (e.g. membrane keypads, touch panels, visual instrument panels). In accordance with this wide range of application, a variety of specially designed screen-printable adhesives with different types of curing mechanisms are available to the market. Acrylic systems, water-based dispersion systems, hot melt adhesives and UV-curable adhesives are the predominant systems available to screen printing technology. The selection of a certain type of adhesive should be based on criteria such as the chemical resistance against the electrolyte that needs to be enclosed inside the battery, the requirements regarding tensile forces or peeling forces the bond has to withstand, and the type of curing mechanism that can be applied according to the properties of the carrier substrate.

Screen printing of dispersion-based adhesives and hot melt adhesives has been successfully demonstrated as an encapsulation method for screen-printed NiMH and zinc-carbon batteries [6]. Printed adhesive layers must be designed to be thinner than the overall height of the electrode layers and the separator to assure proper contact of the electrodes with the separator and the ionic conductive electrolyte after sealing is finished. Depending on the type of adhesive printed, contact pressure has to be applied to the adhesive surface to ensure proper adhesion. To avoid air pockets within the battery, it is necessary to ensure that excessive ambient air is removed before the last side of the adhesive layer is sealed. Thicker layers of printed adhesives or the existence of air pockets facilitate delamination of the electrodes from the separator resulting in a sharp voltage drop with consequent failure of the battery. Adhesive layer thickness needs to be controlled by adjusting the theoretical ink volume of the screen printing meshes to the calculated layer thickness. It should be considered that the layer thickness of the printed adhesive and the applied contact pressure directly influence resistance of the bond to the application of peel or shear forces. Preliminary tests to determine actual printed-layer thickness and resistance to the application of peel or shear forces are necessary and must be performed for each combination of adhesive and substrate in laboratory tests before larger printing runs are realized. Printing adhesives are suitable for the encapsulation of batteries with parallel and sandwich-type architectures in single and multicell arrangements. The resulting barrier capability of the bonding has to be determined in a series of experiments in environmental conditions at different temperatures and moisture ranges, allowing calculations of shelf-life and service life of the manufactured batteries.

7.4.6.2 Contact Heat Sealing

Contact heat sealing is a joining technology basically known from the area of food packaging and is applied for sealing tubular bags that have been filled with a particular product. Contact heat sealing is realized by attaching both sides of two heat-sealable polymer layers facing one another, with a heat jaw pressed against the outer layers of the substrates. Heat is conducted from the outer surfaces of the substrates to the inner layers, and the bonded surfaces are heated to the temperature needed for the layers to soften and melt. The bonding of the two substrates is completed by the application of pressure, forming a characteristic seam, with a subsequent cooling down after the pressure of the heat jaw is released.

The application of this sealing technique requires carrier substrates with heat-sealable thermoplastic surface layers having low melting points, which enables welding temperatures of less than 140 °C. Surface layers made of PE or PP with layer thicknesses of at least 20–30 µm are suitable for contact heat sealing [30]. While this technique allows excellent encapsulation of printed batteries by the direct bonding of the thermoplastic layers of the carrier substrates, it must be considered that even a short period of contact between the substrate and the heat jaw causes considerable heat input to the electrochemical system inside the battery. Reliable heat sealing without damaging the printed lead-out structures of the terminals or the carrier substrate by tearing at the heat-sealed edge depends on controlling the temperature, applied

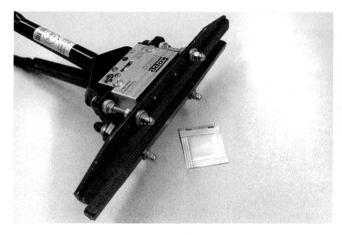

Figure 7.19 Manually realized encapsulation of a printed zinc-based battery with a handsealer.

pressure and pressing time of the heat jaws. In the area of food packaging, pressing times of usually less than 0.5 s are applied to achieve proper melting of the thermoplastic layers of the substrate, resulting in excellent encapsulation, with the barrier capability being defined by the particular material composition of the substrate. While manual contact heat sealing can be realized by use of handsealers (see Figure 7.19), the significant investment costs necessary for the sealing unit and individual sealing-tools for the heat jaws when the layouts of the batteries to be encapsulated are changed represents a major drawback of this technology. Furthermore, process parameters must be validated according to the printing speed and individual material properties to prevent thermal degradation of the electrolyte or the carrier substrate.

7.4.6.3 Ultrasonic Welding

Another approach to sealing printed batteries is the application of ultrasonic welding, which is also used in the field of food packaging as well as for packaging of many other consumer products, such as electronic devices (e.g. blister packaging). Ultrasonic welding is an industrial joining technology whereby mechanical sound waves in a thermoplastic are converted into frictional heat which is locally applied to substrates held together under pressure, creating material bonding at the molecular level and resulting in a solid-state seam. Ultrasonic acoustic vibrations in frequency ranges between 20 kHz and 70 kHz are used for ultrasonic welding in the area of packaging materials at advantageous sealing times of between 80 ms and 200 ms, enabling short cycle times [31]. Figure 7.20 shows details of the components of an ultrasonic welding machine with the booster, sonotrode (active tool) and contoured anvil (passive tool)

(a)　　　　　　　　　　　　　　　　(b)

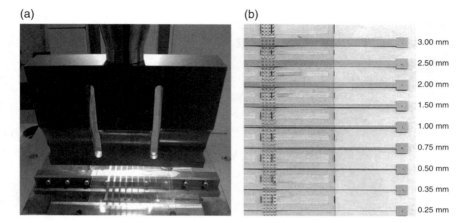

Figure 7.20 (a) Ultrasonic welding of two PET films with printed silver tracks being crossed by the seam. Partial damage to the printed silver tracks simulating lead-outs of printed batteries in variable line widths is clearly visible in (b).

defining the shape of the seam. In Figure 7.20 two PET films are welded with printed silver tracks of different widths being crossed by the seam. Electrical conduction of the printed silver tracks was measured before and after welding of the substrates, allowing determination of possible damage to the tracks by the generation of the seam.

This technology allows the joining of dissimilar materials, which makes ultrasonic welding attractive for the encapsulation of printed batteries consisting of different types of carrier substrates (e.g. printed zinc-air batteries). Another advantage is the local generation of the heat needed for melting of the thermoplastics just inside the sealing zone, preventing the carrier substrate and the particular electrochemistry of the battery from being thermally degraded or damaged. The mechanical vibrations that are introduced into the sealing zone prevent any contamination of the melt and thus the bonding by driving the contents out of the sealing zone [31]. This is a benefit when batteries with aqueous electrolytes are encapsulated and assures flawless seams. Process parameters must be validated for each individual substrate with its particular material composition and layer thickness to achieve tight and strong bonding of the substrates without destruction of the lower layers, which is necessary for physical stability and barrier capability. Once validated, the settings of the process parameters can be digitally saved in a settings file allowing high reproducibility of this joining technology. From the materials standpoint, ultrasonic welding appears to be the ideal sealing technology for printed batteries as excellent bonding of the substrates can be achieved. Evaluation of process parameters is essential to prevent damage to the lead-outs of the terminals crossed by the seam, which may result in the electrochemistry of the manufactured battery being electrically inaccessible. The design of the sonotrode has to be adapted to the particular layout of the printed battery as well as to the desired carrier substrate, which causes high tooling costs compared to the printing of adhesives. This technology can be integrated in production lines like roll-to-roll screen printing machines but it is only an economically astute investment when high volumes of printed batteries with identical layouts are produced.

7.4.7 Conclusion

The interaction of many materials and processes is vital and determines the level of vertical integration in the field of printed batteries. Excellent interaction and process capability of all materials and process steps are mandatory requirements for high-volume printing of batteries with reliable electrical performance characteristics. Table 7.4 shows an evaluation matrix of the technology-readiness levels of individual components involved in the manufacture of printed NiMH and zinc-based batteries at ambient environment.

Table 7.4 Evaluation matrix of the technology-readiness levels of the individual components of printed NiMH and zinc-based batteries.

Electrochemical system	Carrier substrate	Current collectors	Electrode printing pastes	Electrolyte printing pastes	Encapsulation technology
Nickel-metal hydride	2	3	2	2	2
Zinc-carbon, alkaline	2	3	3	2	2
Zinc-carbon, zinc chloride	4	4	4	3	3
Zinc-air, primary	2	3	3	1	1
Zinc-air, secondary	1	2	2	2	1

Scale:

1: materials at early experimental stage

2: materials enable production of individual prototypes in laboratory environment

3: material developments enable small-scale series of prototypes with intermittent production

4: advanced materials allow continuous production on industrial scale.

The evaluated criteria in Table 7.4 show that significant progress has been made in the field of printed zinc-carbon batteries, with zinc chloride as electrolyte constituting the most technologically advanced printed battery system thus far. Excellent interaction and process capability of the individual components enable production of zinc-carbon batteries in industrial quantities with sheet-fed or roll-to-roll screen printing machines. Despite being a promising technology with a high level of vertical integration, a major drawback that has impeded industrial manufacture of printed batteries until now is represented by the deficit of automated in-line quality assurance systems with respect to electrical functionality and encapsulation.

7.5 Industrial Applications and Combination With Other Flexible Electronic Devices

During the past decade many market analyses with the aim of identifying suitable applications for printed batteries have been performed. Various criteria need to be taken into account before devices powered by printed and flexible batteries are designed:

- primary or rechargeable battery
- separate battery or *"printegration"*
- costly but high-quality application or low-cost application.

These aspects need to be evaluated carefully before a type of battery is selected for each particular application. From an economic point of view, these aspects determine whether the desired application is feasible and reasonable or not. Potential applications can be located in various market segments ranging from low cost, like advertising, gaming or RFID, over medium cost, like monitoring of environmental conditions (e.g. temperature, humidity, gas composition), up to high cost, like health-monitoring devices (e.g. blood glucose, lactate). Thus far, two areas of application have been identified as having the most promising outlook for widespread implementation of printed batteries in high quantities.

7.5.1 Self-powered Temperature Loggers

These objects consist of a printed circuit, an antenna, a printed battery and a silicon chip. The attached silicon chip provides all functionalities necessary: temperature sensor, analog-to-digital converter (ADC), µController and near-field communication data transmission. Initially, the circuitry and antenna are printed with a highly conductive silver printing paste on the substrate. The battery can be hybridized or directly printed on the substrate in subsequent printing steps. Finally, the chip is mounted by pick-and-place technology.

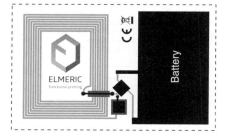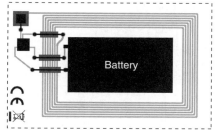

Figure 7.21 Layout variations of temperature loggers with individual positioning of the antenna and battery to optimize area usage.

Printing technologies enable the possibility of producing smart labels that combine sensor functionality and temperature-data-logging capabilities in a cost-effective way. These printed temperature loggers (T-Loggers) can be tailored to customer-specific requirements. Furthermore, the small and particularly thin credit card-sized form factor enables temperature logging at item level and not just at pallet or package level. Figure 7.21 shows two possible layout variations for temperature loggers, both with the same overall footprint but with different antenna and battery area usage. The adaptability of the system components, e.g. the capacity or the nominal voltage of the battery, which can be modified by printing series- or parallel-connections, enable adaptation of the T-Logger to product-specific requirements.

The manufactured smart labels can be programmed and controlled to record a temperature history at defined intervals via the corresponding smartphone app. The data is stored in the attached silicon chip and can be read out by a smartphone or other device with NFC capability anytime and anywhere throughout the supply chain. Once the data set has been read out by a mobile device, it can be directly uploaded to the cloud for further data processing.

The charts presented in Figure 7.22 and Figure 7.23 show exemplary data protocols recorded in a freezer and a heating cabinet and thus demonstrate the typical temperature limits of these printed battery-powered NFC logging devices.

The charts presented in Figures 7.22 and 7.23 demonstrate clearly that the battery voltage is directly related to the occurring temperatures. The battery voltage decreases in cold environments and increases when the battery is heated. This makes the battery the active element of such self-powered logging devices by which the operating temperature range is defined. Typical operating temperatures range between −20 °C and 50 °C. The temperature accuracy and the temperature resolution are defined by the type of sensor integrated into the silicon chip. The memory size and thus the recording period of the T-Logger are defined by the specifications of the silicon chip that is selected and attached.

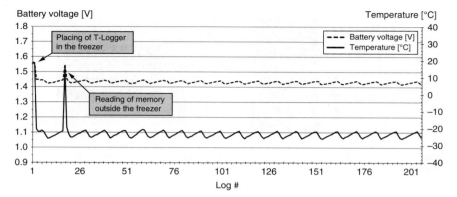

Figure 7.22 Exemplary data protocol of a functional test of a temperature logger with a printegrated zinc-carbon battery (ZnCl, single cell) which was stored in a freezer.

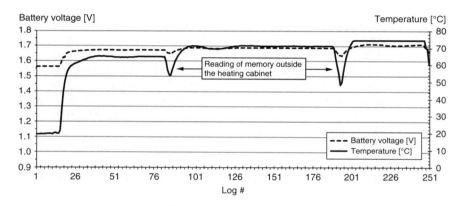

Figure 7.23 Exemplary data protocol of a printed zinc-carbon battery-powered temperature logger (ZnCl, single cell) which was stored in a heating cabinet at elevated temperatures for functional performance testing.

7.5.2 Smart Packaging Devices

Devices for smart packaging are predicted to have a broad range of fields of application. They can be used for advertising purposes, such as talking packaging, for illuminated boxes, such as high-value packaging, for noble beverages, and for logistical purposes in a similar way to RFID transponders.

The specified applications involve smart objects consisting of several printed components that are combined to provide the desired functionality. The printed battery can be hybridized or directly printed on the particular object, which is referred to *printegration*. From the cost and production point of view it is important that all components can be printed on the same substrate.

Combinations of components printed on different types of substrate increase the complexity of the production process by the need for connecting elements. Furthermore, any process that cannot be integrated into a continuous production process decreases production speed resulting in increased manufacturing costs as well as increased unit costs of the final product.

Basically, an increased number of components also increases the degree of complexity and complicates the transfer of the production process to an industrial scale. Growing complexity of products also reduces production yield because a higher number of components may potentially fail during the manufacturing process. Components that cannot be realized by printing technologies must be attached by other technologies like pick-and-place. In general, the combination of printed batteries with other printed objects opens entirely new areas of applications and access to large and predominantly self-perpetuating markets like the packaging or medical sectors.

7.6   Industrial Perspective on Printed Batteries

Printed batteries must be seen in the context of traditionally manufactured batteries. Conventional batteries are produced with state-of-the-art (SOA) technology and machine techniques. The casings are made separately of metal or plastic and the electrodes are produced as pellets, pastes or foils. The separator is made of fiber fleece or microporous membranes made of plastics and the electrolyte is usually in a liquid state. Accordingly, the production equipment is built of highly advanced machinery that automatically mounts and handles objects at high process speeds and in short cycle times. The tooling costs are significant and the design flexibility is low. The mechanical limit of the electrode thickness is in the order of 1 mm.

In contrast to conventional batteries the production of printed smart objects can be realized on standard sheet-to-sheet or roll-to-roll printing machines that are widely used for the production of labels or packaging material. Many process steps are well known and layout changes can be easily performed at comparatively low costs. Printed batteries are only applicable for overall thicknesses of up to 1 mm to ensure the advantage of their thin and flexible character.

7.6.1   Competition with Conventional Batteries

For the reasons given above, printed batteries are at present unable to compete with conventional batteries. Conventional batteries reveal higher energy densities due to better volume ratio of the active masses to the passive components. While the volume of the casings is comparatively low, most of the overall volume of the batteries can be captured by the active masses of the electrodes. Applications like portable electronic devices (e.g. smartphones, tablets, notebooks) require high energy and high power and thus will be equipped with SOA batteries into the future. The advantage of printed batteries is the provision of thin and flexible power sources with overall thickness of less than 1 mm, thus opening new markets that are inaccessible to SOA batteries. The energy densities of printed batteries are comparatively low due to the adverse volume ratio of active masses and passive materials. For example, the active masses of printed $ZnMnO_2$ batteries are represented by the zinc anode with typical layer thickness of approximately 75 μm and the manganese dioxide cathode layer with typical thickness of about 100 μm. Components that are passive as regards the electrochemical cell reactions are represented by the particular carrier substrate with usual thickness of around 120 μm, the current collectors (10 μm), the passivation layer made of carbon (35 μm) as well as the electrolyte/separator combination, exhibiting a characteristic thickness of about 20 μm if printed. The specified values are for coplanar batteries and need to be doubled for sandwich-type batteries in stack configuration. These aspects exclude printed batteries from competing with conventional manufactured batteries with standardized layouts. But their thin and flexible design makes printed batteries the product of choice when flexible and non-standardized power sources are needed.

The production of conventional batteries as a separate item is a benefit of SOA batteries. The separate production of the power source allows 100% monitoring of important quality parameters of each individual battery. Usually quality control comprises measurement of the open-circuit voltage (OCV) and closed circuit voltage (CCV). The separate production of the battery and the particular application or product is economical and highly standardized.

Quality assurance is also possible for printed batteries that are used for hybridization and connected afterwards to the particular product on a separate assembly line. Thus, testing of the functionality of the batteries can be performed after the final process step of battery manufacture. Failed samples can be marked and excluded from further production steps. Performance checks can also be performed on *printegrated* batteries at the end of the production. In this case quality assurance covers function testing of the complete application. If a malfunction is detected, the complete product has to be rejected; this causes high expense if quality defects are not detected at an early stage of the manufacturing process.

Higher voltages can be realized by printing series connections of the particular electrochemical system. The nominal voltage of a single cell is then multiplied by the number of cells printed. Figure 7.24 shows a fully printed series connection of an alkaline zinc-carbon battery realized by screen printing of 20 cells and a printed electrolyte/separator combination.

Series connections of primary battery systems can be printed without problems. The resulting capacity of the series connection is equal to the lowest

Figure 7.24 Battery demonstration of a screen-printed series connection of a primary zinc-carbon battery (alkaline) with a nominal voltage of 30 V consisting of 20 cells. Manufacturing took place during the term of the EU-funded project "Flexibility", contract number: FP7 – 287568.

capacity of the single cells. Printing of rechargeable battery systems may cause problems by an uneven state of charge (SOC) of the single cells. Then the lowest cell can be deeply discharged and driven into reverse followed by complete destruction of the particular cell. Protective circuitry is available and provides a charge balancing between the single cells of large SOA batteries in multicell configurations. The proper function of this protective circuitry requires a battery management system (BMS); otherwise these circuits are inoperable. Precise and repeatable printing of the electrode layers and continuous quality assurance with respect to layer thickness and determination of the deposited active masses of the electrodes avoids the manufacture of cells with uneven states of charge and thus deep discharges of individual cells during their lifecycle. Market development for printed batteries strongly depends on future market trends for smart objects. Those market trends are driven by customer demands but even more by political directives.

7.6.2 Cold Chain Monitoring

Cold chain monitoring is a reasonable and promising application for T-Loggers powered by printed batteries for an individual recording period. Up to now it has been important to control and track the cold chain of products during transport. Fields of application for T-Loggers range from consumer products of daily life, such as frozen food, to highly sensitive areas like the monitoring of transport conditions for blood reserves, when a certain temperature cannot be exceeded. The existing measurement equipment is too expensive and makes controlling and tracking of temperatures of individual products impossible. High availability of self-powered and fully printed T-Loggers at acceptable prices would initiate increasing demand, with more and more transported goods being equipped with these monitoring devices.

7.6.3 Health-monitoring Devices

Demographic change in populations has led to an ever-increasing number of older people who need to be supported in their daily lives. Growing demand for affordable health-monitoring equipment for the surveillance of blood pressure, heart rate, breathing rate, blood-glucose levels and many more values is noted. The development of flexible and self-powered monitoring devices enables improvements in the care and treatment of older people. Existing monitoring devices are powered by small SOA batteries and thus are inconvenient to carry due to their rigid construction. Printed thin and flexible devices would make the lives of patients and older people much easier through their adaptable designs. Future trends in the medical sector must be observed carefully so that the industry is prepared when the need for printed smart objects increases.

7.7 Conclusion

The large number of current research activities in the area of printed batteries indicates the continuing interest in getting this technology ready for the market. Innovative and directly *printegrated* applications like self-powered monitoring devices are predicted to be the most likely to enter the market in high quantities within the next few years. Compared to conventional battery manufacturing methods, printing technologies are characterized as cost-effective and highly flexible technologies and thus as economic solutions for both small and larger batch sizes of printed batteries and other functionalities with individual layouts and specifications.

Recently, the first T-Loggers powered by fully screen-printed primary batteries for cold chain monitoring of logistic services were reported. Extending the product range for printed battery applications requires further research in the areas of material sciences, functional printing technologies and packaging technologies. With more primary and secondary electrochemical systems being accessible by printing technologies, the number of potential applications and devices with higher demands on battery voltages and power outputs will subsequently increase. Further improvements in printed batteries with respect to electrical performance parameters, life-cycle and reliability are of the highest importance as well.

For the moment, printed electronics, especially printed batteries, are considered more likely to be a technology-push. It is essential to motivate and persuade customers to generate a significant market-pull for printed and flexible electronics devices. Progress will be the greatest when technology-push and market-pull have nearly reached a balanced level. This condition is foreseen to accelerate the development of further innovative products with integrated printed power sources.

References

1 Kang, H., Park, H., Park, Y., Jung, M., Kim, B.C., Wallace, G. *et al.* (2014) Fully roll-to-roll gravure printable wireless (13.56 MHz) sensor-signage tags for smart packaging. *Scientific Reports* **4**, 5387.

2 Tobjork, D., Osterbacka, R. (2011) Paper electronics. *Adv. Mater.* **23**, 1935–1961.

3 Li, Y., Zhang, L., Li, M., Pan, Z., Li, D. (2012) A disposable biosensor based on immobilization of laccase with silica spheres on the MWCNTs-doped screen-printed electrode. *Chemistry Central J.* **6(1)**, 103.

4 Bandodkar, A.J., Jia, W., Wang, J. (2015) Tattoo-based wearable electrochemical devices: a review. *Electroanalysis* **27**, 562.

5 Reddy, T.B. (2011) *Linden's Handbook of Batteries*, 4th edn, McGraw-Hill, New York, NY.

6 Wendler, M., Hübner, G., Krebs, M. (2010) Screen printing of thin, flexible primary and secondary batteries. In N. Enlund and IARIGAI (eds) *Advances in Printing and Media Technology, Vol. XXXVII: Proceedings of the 37th international research conference of IARIGAI, Montreal, Canada, September 2010, 12–15,* International Association of Research Organisations for the Information Media and Graphic Arts Industries (IARIGAI), Darmstadt, 303–312.

7 Suren, S., Kheawhom, S. (2016) Development of a high energy density flexible zinc-air battery. *J. Electrochem. Soc.* **163(6)**, A846–A850.

8 Gaikwad, A.M., Arias, A.C., Steingart, D.A. (2015) Recent progress on printed flexible batteries: mechanical challenges, printing technologies, and future prospects. *Energy Technology* **3**, 305–327.

9 Madej, E., Espig, M., Baumann, R.R., Schuhmann, W., La Mantia, F. (2014) Optimization of primary printed batteries based on Zn/MnO2. *J. Power Sources* **261**, 356–362.

10 Gaikwad, A.M., Whiting, G.L., Steingart, D.A., Arias, A.C. (2011) Highly flexible, printed alkaline batteries based on mesh-embedded electrodes. *Adv. Mater. (Deerfield Beach, Fla.)* **23**, 3251–3255.

11 Choi, M.G., Kim, K.M., Lee, Y.-G. (2010) Design of 1.5 V thin and flexible primary batteries for battery-assisted passive (BAP) radio frequency identification (RFID) tag. *Current Applied Physics* **10(4)**, e92–e96.

12 Hilder, M., Winther-Jensen, B., Clark, N.B. (2009) Paper-based, printed zinc–air battery. *J. Power Sources* **194** (2), 1135–1141.

13 Willmann, J., Stocker, D., Dörsam, E. (2014) Characteristics and evaluation criteria of substrate-based manufacturing. Is roll-to-roll the best solution for printed electronics? *Organic Electronics* **15**, 1631–1640.

14 Heraeus Deutschland GmbH & Co. KG (2016) SOL9631 Series: New Generation PER C Front-Side Silver Paste.

15 Kipphan, H. (2001) *Handbook of Print Media: Technologies and production methods*, Springer Electronic Media, Berlin.

16 Lorenz, A., Senne, A., Rohde, J., Kroh, S., Wittenberg, M., Krüger, K. et al. (2015) Evaluation of flexographic printing technology for multi-busbar solar cells. *Energy Procedia* **67**, 126–137.

17 Wang, Z., Winslow, R., Madan, D., Wright, P.K., Evans, J.W., Keif, M. et al. (2014) Development of MnO2 cathode inks for flexographically printed rechargeable zinc-based battery. *J. Power Sources* **268**, 246–254.

18 Wang, Z. (2013) Flexographically printed rechargeable zinc-based battery for grid energy storage. University of California. PhD thesis.

19 Galus, M. (2009) Tiefdruck. In M. Lake and M. Boes (eds) *Oberflächentechnik in der Kunststoffverarbeitung: Vorbehandeln, Beschichten, Funktionalisieren und Kennzeichnen von Kunststoffoberflächen*, Hanser, München, 212–235.

20 Schneider, A., Traut, N., Hamburger, M. (2014) Analysis and optimization of relevant parameters of blade coating and gravure printing processes for the fabrication of highly efficient organic solar cells. *Solar Energy Materials and Solar Cells* **126**, 149–154.

21 Park, J.D., Lim, S., Kim, H. (2015) Patterned silver nanowires using the gravure printing process for flexible applications. *Thin Solid Films* **586**, 70–75.

22 Secor, E.B., Lim, S., Zhang, H., Frisbie, C.D., Francis, L.F., Hersam, M.C. (2014) Gravure printing of graphene for large-area flexible electronics. *Adv. Mater. (Deerfield Beach, Fla.)* **26** (26), 4533–4537.

23 Steiner, E. (2007) *Printed Electronics: Fundamentals of Polytronics*, lecture notes, Stuttgart.

24 Sefar AG Printing Division (2009) *Handbook for Screen Printers*, Thal.

25 Erath, D., Filipović, A., Retzlaff, M., Goetz, A.K., Clement, F., Biro, D. et al. (2010) Advanced screen printing technique for high definition front side metallization of crystalline silicon solar cells. *Solar Energy Materials and Solar Cells* **94(1)**, 57–61.

26 Pfletschinger, M., Petersen, I. (2011) Elektronik auf Folien drucken: Material- und Fertigungskonzepte für Folienantennen. *ATZ Elektron* **6(4)**, 46–51.

27 Varadharaj, E.K., Jampana, N. (2015) Studies on carbon mediated paste screen printed sensors for blood glucose sensing application. *ECS Transactions* **66(38)**, 23–33.

28 Wendler, M., Steiner, E., Claypole, T.C., Krebs, M. (2014) Performance optimization of fully printed primary (ZnMnO2) and secondary (NiMH) batteries. *J. Print Media Technol. Research* **III(4)**, 241–251.

29 Baggetto, L., Niessen, R.A.H., Roozeboom, F., Notten, P.H.L. (2008) High energy density all-solid-state batteries: a challenging concept towards 3D integration. *Adv. Funct. Mater.* **18**, 1057–1066.

30 Hishinuma, K., Miyagawa, H. (2009) *Heat Sealing Technology and Engineering for Packaging: Principles and Applications*, DEStech Publications, Inc., Lancaster, PA.

31 Fischer, T. (2009) Packaging with ultrasonics. *Kunststoffe* **11(8)**, 51–55.

8

Open Questions, Challenges and Outlook

Carlos Miguel Costa[1,2], Juliana Oliveira[1] and Senentxu Lanceros-Méndez[1,3]

[1] Center of Physics, University of Minho, Gualtar campus, Braga, Portugal
[2] Center of Chemistry, University of Minho, Gualtar campus, Braga, Portugal
[3] BCMaterials, Basque Center for Materials, Applications and Nanostructures, Spain

The battery proposed by Volta in 1800 showed the possibility for storing electrical energy using chemical reactions. Today, it remains a technological highlight and a revolutionary discovery. Throughout the last century, many types of batteries, including NiCd: nickel-cadmium, NiMH: nickel-metal hydride, and, in particular, Li-lithium, have been developed, representing relevant technological advances [1].

In light of advances in printed electronics, the junction between battery technologies and printing techniques has resulted in printed batteries, which represent an alternative to conventional batteries in many applications, such as wearable and portable devices.

As we have seen in this book, the main printed battery types that have been developed are Zn/MnO_2 and lithium-ion, and one of the most important attributes of printed batteries is their being customizable for small, thin, light-weight devices, an area where a promising future is expected.

Figure 8.1 summarizes the main characteristics of printed batteries; the most relevant challenges to be tackled are to increase current and energy density of the battery.

Cost, safety and ease of integration are also challenges that require particular attention.

These challenges are important not only for current applications of printed batteries such as RFID devices, smart cards, sensors and wearable and medical devices, but also for smart objects, increasingly demanded in the scope of the growing "Internet of Things, IoT" concept [2] (Figure 8.2).

Printed Batteries: Materials, Technologies and Applications, First Edition.
Edited by Senentxu Lanceros-Méndez and Carlos Miguel Costa.
© 2018 John Wiley & Sons Ltd. Published 2018 by John Wiley & Sons Ltd.

Figure 8.1 Main characteristics of printed batteries.

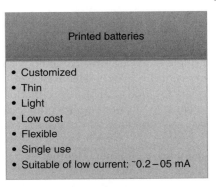

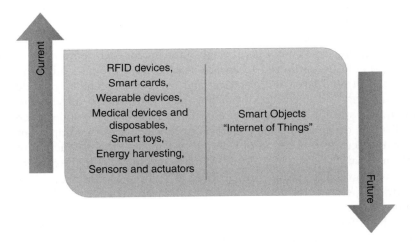

Figure 8.2 Current and future applications of printed batteries.

Thus, the continuous and increasing advances in the areas of smart and functional materials technology strongly support the Internet of Things to become a ubiquitous network where everything is connected to the Internet [3, 4]. This new paradigm brings new functionalities to the service and production sectors, among others, allowing increasing efficiency, speeded up processes and the incorporation of more complex and flexible organizational systems [5, 6].

In the rapidly growing field of the Internet of Things, smart things/objects require batteries that are flexible, light and cheap, with rapid charging and a large variety of durations; thus, printed batteries are among the most suitable solutions for many applications.

For the fabrication of printed batteries, the choice of printing technique requires particular attention, taking into account the ink properties for each component of the battery, with inkjet, screen and spray printing being the most frequently used printing techniques.

Recently, 3D printing technologies are being applied to the development of printed batteries. This technology has been successfully applied to all components of the battery (electrodes and separator/electrolyte) and can be appropriate for industrial-scale production [7].

Thus, 3D-printed interdigitated electrodes have been developed, leading to fast insertion kinetics of ions at increasing specific currents [8].

Further, 3D Li-ion batteries based on $LiMn_{0.21}Fe_{0.79}PO_4$ (LMFP) nanocrystal cathodes exhibited better long-term stability than the coated-electrode Li-ion battery, also showing high charge/discharge rate and high capacity in comparison to conventional batteries [9]. Similarly, 3D printable Li-ion battery electrolyte has been produced, yielding a high-performance printed electrode membrane assembly [10].

Control of the ink properties is an essential step in the successful development of printing devices. Thus, the rheology of the ink for each component of the battery should be optimized to ensure reliable flow, promote adhesion between each printed layer and electrical and mechanical stability, provide structural integrity and avoid cracking during the drying process. Inks have to be tailored for a specific printing technology in terms of viscosity and surface tension. Further, the substrate and its intrinsic properties are other important factors that affect the application of printing systems and the production of printable materials [11]. The majority of printable materials are in the form of solutions, which require specific properties to allow proper printing, such as nanoparticle dispersion or a moderate and suitable level of chemical and physical stabilities [12].

Among the different components of printed batteries, the one that requires special attention is the separator/electrolyte, due to its high ionic conductivity and electrochemical, thermal and mechanical stability.

This battery component needs further development and the incorporation of ionic liquids in the polymer solutions appears a promising solution for improving the actual performance of printable separators/electrolytes. Further, novel engineered (nano)materials for the different components are continuously being developed, allowing further increase in printed battery energy density.

It is important to note that printed batteries also allow the development of new, unconventional geometries, allowing better integration into planar, flexible devices or devices with complex geometries.

The most relevant issues to be overcome in order to increase battery performance are related to the understanding and optimization of the interfaces of the different printed layers. This issue must be solved, not only to increase battery performance, but also to improve scalability of printed batteries and reduce production costs.

The aforementioned primary challenges, among others, will drive the main research activity in this field, allowing printed batteries to move forward and to provide a new generation of wearables, portable devices and smart solutions for the Internet of Things, thus powering a lighter, friendlier and more flexible future.

Acknowledgements

This work was supported by the Portuguese Foundation for Science and Technology (FCT) in the framework of Strategic Funding UID/FIS/04650/2013, project PTDC/CTM-ENE/5387/2014 and grants SFRH/BD/98219/2013 (J.O.) and SFRH/BPD/112547/2015 (C.M.C.). The authors thank the Basque Government Industry Department under the ELKARTEK Program for its financial support.

References

1 Scrosati, B., Abraham, K.M., van Schalkwijk, W.A., Hassoun, J. (2013) *Lithium Batteries: Advanced Technologies and Applications*, John Wiley & Sons, Inc., Hoboken, NJ.

2 Greengard, S. (2015) *The Internet of Things*, MIT Press.

3 Gubbi, J., Buyya, R., Marusic, S., Palaniswami, M. (2013) Internet of Things (IoT): a vision, architectural elements, and future directions. *Future Generation Computer Systems* **29**, 1645–1660.

4 Atzori, L., Iera, A., Morabito, G. (2010) The Internet of Things: a survey. *Computer Networks* **54**, 2787–2805.

5 Granjal, J., Monteiro, E., Sa Silva, J. (2015) Security for the internet of things: a survey of existing protocols and open research issues. *IEEE Communications Surveys and Tutorials* **17**, 1294–1312.

6 Bandyopadhyay, D., Sen, J. (2011) Internet of things: applications and challenges in technology and standardization, *Wireless Personal Communications* **58(1)**, 49–69.

7 Tian, X., Jin, J., Yuan, S., Chua, C.K., Tor, S.B., Zhou, K. (2017) Emerging 3D-printed electrochemical energy storage devices: a critical review. *Adv. Energy Mater.* 1700127-n/a.

8 Fu, K., Wang, Y., Yan, C., Yao, Y., Chen, Y., Dai, J. *et al.* (2016) Graphene oxide-based electrode inks for 3D-printed lithium-ion batteries. *Adv. Mater.* **28**, 2587–2594.

9 Hu, J., Jiang, Y., Cui, S., Duan, Y., Liu, T., Guo, H. *et al.* (2016) 3D-printed cathodes of LiMn1–xFexPO4 nanocrystals achieve both ultrahigh rate and high capacity for advanced lithium-ion battery. *Adv. Energy Mater.* **6(18)**, 1600856-n/a.

10 Blake, A.J., Kohlmeyer, R.R., Hardin, J.O., Carmona, E.A., Maruyama, B., Berrigan, J.D. *et al.* (2017) 3D printable ceramic–polymer electrolytes for flexible high-performance li-ion batteries with enhanced thermal stability. *Adv. Energy Mater.* 1602920-n/a.

11 Hoffman, J., Hwang, S., Ortega, A., Kim, N.S., Moon, K.S. (2013) The standardization of printable materials and direct writing systems. *J. Electronic Packaging, Transactions of the ASME* **135(1)**.

12 Khan, S., Lorenzelli, L., Dahiya, R.S. (2015) Technologies for printing sensors and electronics over large flexible substrates: a review. *IEEE Sens. J.* **15(6)**, 3164–3185.

Index

Printed Batteries: Materials, Technologies and Applications, First Edition.
Edited by Senentxu Lanceros-Méndez and Carlos Miguel Costa.
© 2018 John Wiley & Sons Ltd. Published 2018 by John Wiley & Sons Ltd.